AUSTRALIAN COCKATOOS

DEDICATION

It is with sadness that I dedicate this book
to my friend, co-author and master aviculturist,
Bob Lynn who did not live to savour the
fruits of our labours

AUSTRALIAN COCKATOOS

Experiences in the Field and Aviary

by

STAN SINDEL
ROBERT LYNN

SINGIL PRESS PTY LTD
P.O. Box 9, Austral, NSW 2171
Australia

ISBN 0 9587727 1 1

Reprinted 1996

Designed and Typeset by
SURREY BEATTY & SONS PTY LIMITED
43 Rickard Road, Chipping Norton, NSW 2170
Australia

Printed in China

Frontispiece
Size comparison of the longest (northern race of the Red-tailed Black Cockatoo, *Calyptorhynchus m. magnificus*) and the smallest (Gang Gang Cockatoo, *Callocephalon fimbriatum*) members of the group *Cacatuini* (typical cockatoos.)

ACKNOWLEDGEMENTS

I am writing these acknowledgements in the first person, due to the death of my co-author, Bob Lynn, since the completion of the manuscript.

First, thanks must go to Bob for his extensive avicultural advice and information on Black Cockatoos.

To all those people, far too numerous to list individually who have provided me directly and indirectly with much of the avicultural and field information contained in these pages, I extend my thanks and gratitude.

Jim Gill, my co-publisher in these avicultural adventures, who is always there with critical and constructive advice as well as proof reading and editing, must be thanked for all this and his concise chapter on cockatoo diseases, yet most importantly for his friendship.

Cyril Laubscher, a very old friend from the U.K., spent a busman's holiday with us last year and kindly devoted much of this time to achieve the superb cockatoo portraits featured in these pages. My sincere thanks Cyril.

Many thanks to Des Spittal, Curator of Currumbin Bird Sanctuary, who kindly consented to write the flattering foreword for this book.

Thanks to all the photographers who supplied photographs for this work. Personal acknowledgements appear in the list of photographers.

Scores of helpful people supplied me with information, contacts and specimens, etc. which enabled me to complete these pages. Thanks to all and particular thanks to Bill Connor, NSW, Russell McAllister, NSW, Dorothy Payne, W.A., Eddy Stevens, W.A., Bob Phillpot, W.A., John De Jose, W.A., Neil Hamilton, W.A., Peter Brindley, W.A., Des Dowling, NSW, Tony Vaughan Johnson, NSW, Lord McAlpine, W.A., Graham Taylor, W.A., Matthew Crane, NSW, John McGrath, NSW, and Glenda Bain, NSW.

A special thank you to Dick Schubot, the Founder of the Avicultural Breeding and Research Centre, Florida, USA, for his kind assistance with nestling photographs of Palm Cockatoos.

My thanks to Don Burke for his help in identification of native flora.

For the comments and corrections as well as for the many hours spent enduring these pages, I extend my sincere thanks to my proof readers Mike Cannon, Therase Wright, Alison Stuart, Marilyn Gill and my wife Jill.

I am also indebted to Stacey Libbesson for the numerous hours spent word processing for me — thanks Stace.

In general I would applaud the journals of the avicultural societies of Australia, in particular "Australian Aviculture" and "Birdkeeping in Australia" as well as the ornithological journals and newsletters of the Royal Australian Ornithological Union and the Bird Observers Club of Australia, all of which in their own particular field, provide so much information to bird people everywhere.

Once again I must thank Ivor Beatty for his advice with the production and his firm for the printing of this book.

TABLE OF CONTENTS

LIST OF PLATES

Photographers — Bob Lynn: 1A, 6D, 6E; Stan Sindel: 1B-F,6A-C,9A-F, 12C-D, 13A-E, 16A-C, 16E, 17, 18A-E, 20A-D, 21A-E, 22A-C, 22E, 24A-C, 26A-E, 28A-E, 31A-E, 33A-E; Cyril Laubscher: 2, 4, 5, 7, 8, 10, 11, 12A,14, 15, 19, 23, 25, 27, 30, 32, 34, 36A-F, 37A-F; Max Lawrence: 3A; Aviculture Breeding and Research Institute: 3B-E; Jim Gill: 24, 35; Lance J. Ruting: 29; G. Miles: 12B; Gordon Dosser: 22D.

FOREWORD

Every so often a publication appears which is a milestone in its field. Such is this book by Stan Sindel and Robert (Bob) Lynn.

It is regretted that Bob did not live to see this book published and appreciate the impact that it will have upon aviculture in Australia and overseas. As Stan mentions in the introduction there are many world firsts achieved by Bob, and his wealth of knowledge in his particular sphere was immense. Bob will be greatly missed.

One of the most important aspects of aviculture that is very often overlooked is communication. Very little of importance can be achieved anywhere if what we learn is not shared. Stan and Bob are to be congratulated in producing a momentous piece of communication in a field which has fascinated and frustrated many aviculturists worldwide — how do you breed and maintain Australian cockatoos?

In this book the beginner, as well as the experienced aviculturist, will find much useful information to assist him in achieving the dream of all breeders. That dream, which is reached by so few, is to produce healthy, parent reared babies. In a way which is easily understood, Stan and Bob take us through the world of Australian cockatoos, and by listening to the advice given by these highly experienced and respected aviculturists success will be that much nearer.

At no time in the past has there been as much pressure on the wild populations of most species of parrots. A particular problem is evident with all types of black cockatoos. Their declining numbers in the wild are due primarily to habitat destruction and to a lesser extent in collection for the aviculture industry. The major thrust of this book, to encourage the captive propagation of these birds, is therefore to be applauded.

One of the tools of conservation is the captive breeding of restricted species with the eventual aim of releasing surplus animals back into the wild. It is only by encouraging more people to undertake this task and strive to achieve these goals that sufficient numbers of captive birds can be bred. These captive stocks should act as an insurance against the loss of any species in the wild. It is for these reasons that I highly commend Stan and Bob on communicating their many years of experiences.

This wealth of information is abundantly backed up by many photographs and maps covering all known types of Australian cockatoos, including mutations. The only area where some readers may find contention is in the handling of the taxonomy of some species.

Currently there are many changes being made to the taxonomy of all Australian psittacines and it would not come as a surprise if many of the changes suggested here, based on many years of observations in captivity and the wild, became standard.

This book does not only represent a milestone in the literature of aviculture, but also a cornerstone upon which will be developed a vast array of ideas and ideals.

The information contained within these covers will last for years.

Des Spittal,
Curator,
Currumbin Sanctuary,
Queensland,
Australia
1988

INTRODUCTION

During the last few decades we have seen the emergence of worldwide fashions, or should I say crazes, in particular psittacine groups. These are obviously encouraged by the shrinking world boundaries created by modern communication and media networks that influence all aspects of our lives today. Any trend that emerges anywhere, be it political, monetary, dietary, whatever, good or bad, soon becomes a trend of world scale proportions.

Worldwide, in the sixties, Australian parrakeets were the bird to own and the rarer the species the more desirable it was for the aviary collections. The seventies saw the trends move to African Lovebirds with all their newly created colour forms. This worldwide boom was to be the greatest ever seen in aviculture, but like all booms they eventually fade and are followed by the inevitable decline.

What causes these shifts in avicultural interest? No doubt supply and demand is the major influence. We all desire to own something that is rare, then when it is acquired, we set about to propagate it to such an extent that it soon becomes common. Our own modern avicultural techniques and ever improving expertise, together with greed, create these upsurges of interest in particular genera, then just as quickly, their demise.

After the boom years and the inevitable depressions in these various prolific avian groups and the assessment of the monetary gains and eventual losses as the market reached saturation point, thinking aviculturists around the world looked for an answer. The solution was to find desirable avicultural species whose reproductive abilities were limited and whose protection in their country of origin prevented the market being flooded by wild caught birds — Australian Cockatoos were an obvious answer.

Hence the eighties saw the start of the cockatoo era, another fashion that will pass, but far more slowly, for nothing happens quickly with large psittacines. Even with all our modern technology, breeding is still mainly a waiting game.

The pinnacle of aviculture is reached when an aviculturist can provide a pair of birds of a difficult group such as the cockatoo family, with the facilities and dietary requirements necessary to promote the production of healthy, vigorous, parent reared young. It is this ability to be able to sense the needs of a difficult species that places the great aviculturists of the world above the rest of us.

I consider my co-author Robert Lynn (Bob Lynn) to be one of these great aviculturists. Like most of us, Bob's early years in aviculture consisted of experimentation in various parrot species, but by the late 1950's an intense liking, or perhaps obsession is a more appropriate word, had developed for the Black Cockatoo species. The next decade was spent collecting all species and developing his avicultural techniques and the dietary requirements necessary for their propagation.

My first meeting with Bob occurred in the early 1960's when I joined the Avicultural Society of NSW and he was the incumbent president. Over the years we developed a close friendship which enabled me to follow Bob's avicultural achievements with his Black Cockatoos closely, and to have my own interest kindled in this group, although I did not personally keep Black Cockatoos until quite recently.

Bob Lynn's most outstanding avicultural achievements are:

He is the only man in the world to successfully breed all species of Black Cockatoos (all young being parent reared).

He received a world first breeding award for the Yellow-tailed Black Cockatoo *(Calyptorhynchus funereus)* from the Avicultural Society of NSW.

He received a world first breeding award for the Palm Cockatoo *(Probosciger aterrimus)*, also from the Avicultural Society of NSW.

He has the world first documented breeding of the White-tailed Black Cockatoo *(Calyptorhynchus Baudinii latirostris)* which is almost certainly a world first breeding.

Then recently Bob had two beautiful, parent reared young White-tailed Black Cockatoos fledge from the same nest (the normal clutch reared is one) which is yet another world first.

Unfortunately, Bob has never been a person to put pen to paper, although he has kept precise records of his breedings. His information, compiled over 30 years and recorded in these pages, is of immense avicultural significance.

As a birdman Bob has only one failing, if a bird is not generally blackish in colour with a hooked bill he just does not see it. Yet this failing, or obsession if you like, is the very thing that promoted his intense specialisation and his success.

Our continued aim with these specialised handbooks is to provide aviculturists at all levels with further knowledge and a full understanding of their birds, as well as their hobby, in the past, present and perhaps the future. The intention is always to keep it simple and within the range of all our comprehensions, yet to provide all current information and explain the modern techniques used in aviculture today. Any avicultural or ornithological words or terminologies used that may present difficulties to newcomers into aviculture have been defined on their initial use.

Australian cockatoos are a diverse group from the family Psittacidae, subfamily Cacatuinae and members of two distinct groups, the Cacatuini and the Calopsittacini.

The Cacatuini Group contains all genera of cockatoos except one and consists of Probosciger (Palm Cockatoo), Calyptorhynchus (Black Cockatoos), Callocephalon (Gang Gang), Eolophus (Galah), Cacatua (all white and near white species of cockatoos).

The Calopsittacini group contains one genera, Nymphicus, which in turn has only one species *Nymphicus hollandicus* (Cockatiel).

The Cacatuinae are physically distinguished from other psittacines by their crest and the ability to raise or lower it when excited or agitated. Powder down patches are prominent in their feathering and the feather barbs lack the structural requirements to reflect the necessary lights to produce the colours usually seen in psittacines. The Cacatuini group are all large solid birds with large strong beaks and broad square or rounded tails.

Calopsittacini is a small slender bird with a long tapered tail and the ability to lower the normally erected crest.

Probosciger is a monotypic genus (one member) distinguished by long narrow curved crest feathers, large red bare skin cheek patches, exceptionally large mandibles, bare thighs, a feathered cere and sexes are alike.

Calyptorhynchus are large dark coloured birds with banded tails. They have varying length shell-shaped crests, bare eye rings and cere, large strong mandibles and varying degrees of sexual dimorphism (sexes differ in appearance)

Callocephalon is a monotypic genus containing a smaller distinctive cockatoo with an unusual filmy overflowing crest that curves forward, short thick legs, large strong curved beak, covered cere and pronounced sexual dimorphism.

Eolophus is another monotypic genus containing a smaller cockatoo with distinctive pink and grey plummage, short shell-shaped crest, a pronounced naked eye ring, feathered cere, medium sized strong beak and slight sexual dimorphism.

Cacatua is an extensive genus of predominately white or pale coloured cockatoos, which in my opinion requires considerable revision. They range from small to large sized cockatoos with varying shell to fan-shaped crests, short square tails, long wings, large strong beaks, bare eye rings, predominately feathered ceres and sexes are alike in most species, although there is slight sexual dimorphism in some.

Nymphicus is a monotypic genus containing a very small slender cockatoo with a long upright pointed crest, parakeet-like feet, naked cere, small but powerful beak and marked sexual dimorphism.

Representatives of all genera are found in Australia including 13 of the 20 species recognised throughout the world. Of these 13 species, 10 are endemic to Australia.

Australia has often been termed the "Land of Parrots"; perhaps a more suitable terminology would be the "Land of Cockatoos", for any part of the entire Australian continent and Tasmania is home for at least one species of cockatoo. Nowhere else in the world can so many distinct and varied species of cockatoos be seen, all still in good numbers and many in plague proportions.

Unbelievable sights can now be seen in and around Sydney, such as Gang Gangs eating from a bird feeding station in the heart of suburbia; flocks of Galahs, Sulphur-crested Cockatoos and Short-billed Corellas feeding in parks throughout the city and suburbs. Gang Gangs, Yellow-tailed and Glossy Black Cockatoos not only feeding but breeding in the rural suburbs surrounding Sydney.

In just two days of last week I observed the following cockatoo events. On Wednesday morning, after speaking at the Central Coast Avicultural Society Meeting in Gosford the night before, I observed a beautiful pair of Eastern Long-billed Corellas busily digging nut grass bulbs on the bank of the Brisbane Waters at Empire Bay, almost 1000 km outside their normal range. On the opposite shore at Davistown a flock of Short-billed

Corellas were moving noisily along the waters edge, also a similar distance outside their range. At the same time flocks of Galahs and Sulphur-crested Cockatoos could be seen flying further afield.

Then on the Friday afternoon while driving through Parramatta (in the Sydney area) in heavy peak hour traffic, there, adjacent to several multi-storey office blocks, was a flock of over 20 Galahs sitting along power lines. A kilometre further along the road, while waiting for traffic lights to change at a busy intersection, I could see several Sulphur-crested Cockatoos feeding on the seed pods of the introduced Liquid Amber tree, not four metres from the roadway. Then another few kilometres along Pennant Hills Road at least a dozen Short-billed Corellas were congregated in a large Eucalypt. Surely this must be the "Land of Cockatoos".

All Australian cockatoos, with the exception of the Cockatiel present an avicultural challenge. More and more aviculturists are rising to this challenge of successfully maintaining and breeding our cockatoos. Don't expect Cockatoo Culture to be easy; complete dedication and devotion to the cause is necessary for success, but the reward of seeing healthy, active boisterous cockatoos flying noisily to and fro in your aviaries is superb. The absolute ultimate feeling of ecstasy for a cockatoo-culturist is the vision of healthy, robust, parent reared young on the day they fledge.

These are the great yet simple pleasures of life that only successful aviculturists are fortunate enough to savour. The more challenging the species, the greater the sense of achievement when the ambition is fully realised. I hope these pages help you to realise these pleasures.

I personally find all psittacines rewarding avicultural subjects but I must confess that the individualism and distinctive, if not always pleasing, personalities of cockatoos make them my favourite aviary family and the Gang Gang is and has been my ultimate aviary bird from the time I saw my first aviary specimen over 30 years ago.

Finally, I appeal to all people concerned about conservation and interested enough in cockatoos to read these pages, to commence working for the reversal of the detrimental policies, practices and attitudes implemented against many of our cockatoo species by governments, wildlife authorities, farmers and illegal trappers and to ensure the protection of all cockatoos and their habitats.

HOUSING

Historically, cockatoos have been housed in a fantastic array of cages and aviaries, often ornate, but mostly poorly designed. Little or no thought had been given to the well-being or comfort of the occupants by the designer or builder. As these pages will reveal I have in the past been equally guilty of providing inadequate and unsuitable housing for cockatoos.

The remarkable aspect of cockatoo housing is that we continually hear of breeding successes occurring in some appallingly small cage, or should I say flight cage which is a term commonly used outside Australia in lieu of inadequate housing. It appears to be a common occurrence overseas for pairs of cockatoos to breed in cages with a capacity of little more than a cubic metre and the nest boxes fitted externally.

Surely we as responsible aviculturists must provide housing that allows our cockatoos adequate exercise and room to fly at least a few metres. Over the years it has been my experience that if cockatoos are continually denied exercise they became lethargic and obese, which in turn promotes infertility and egg binding.

In Australia, where some of our cockatoo species are in plague proportions, unprotected and readily available throughout the pet trade, beautiful pairs are often purchased cheaply by novice aviculturists and placed in suitable housing in the hope of achieving a successful breeding. Sometimes these pairs are successful but whether or not, as the aviculturist and his collection progresses so the housing for the common cockatoo becomes less suitable, as the birds are moved from their aviaries to make way for more valuable birds. This is a common problem which I feel is best solved by passing these breeding pairs and potential breeding pairs on to other aviculturists so that they too may enjoy the thrill and sense of achievement experienced by all of us when a species is bred for the first time. This must be preferable to the eventual relegation to a cockatoo cage.

Now we come to the question: what is the most suitable aviary to breed cockatoos in? There is no suitable cage. This is a question with numerous answers which must be interpreted and applied to suit your own particular set of circumstances. I shall endeavour to convey what we consider to be the housing requirements for the successful maintenance and propagation of cockatoos. (See Figure 1.)

Housing falls into two categories: conventional aviaries and suspended wire aviaries.

Designing a Conventional Aviary for the Cacatuini Group

When designing a cockatoo aviary the following points should be considered:

1. *Durability*

Due to the destructive chewing ability of most cockatoos robust construction is essential. The use of exposed timber framework or surfaces is totally inadviseable.

2. *Framework*

Aviary frames should ideally be constructed of square or round galvanised steel tubing, which has proven to be long lasting and almost indestructible. There is a variety of jointing methods available. Downey clamps are simple but expensive. Welding gives an excellent finish. In the case of square tubing all manner of joints can be achieved by cutting the four corners of the end of the tube for 25 mm and folding to accept another piece of tube, then pop rivetting or tech screwing the joint.

3. *Walls*

Walls, like the frame, must be of an indestructible nature. Many fine aviaries have their shelters and partitions constructed of brickwork. They are strong and give an excellent appearance, but have the disadvantages of being expensive and of a porous nature, which makes it difficult to clean.

Metal sheeting, particularly colour bond, which does not need to be painted, is durable and has a non-porous, easy to clean surface, may be fixed easily on to a metal framework and is less expensive than brickwork.

Fibro cement sheeting is an unsuitable material for walls and partitions as cockatoos will take great delight in destroying it once they find an edge to start chewing on.

I have used corrugated asbestos cement sheeting for walls in many aviaries housing cockatoos and macaws without problems. This product is no longer manufactured but second-hand sheets can sometimes be obtained very cheaply. Asbestos cement sheeting does have the disadvantage of being porous and hence hard to clean.

4. *Roofing*

Metal sheet roofing is less subject to storm damage than other materials but may require insulation in extreme climates. When roofs are galvanised or painted white, heat is reflected thus reducing temperatures within the aviaries. I currently use a corrugated metal sheeting fixed on top of flat fibro cement sheeting for the roof and I find this greatly reduces the temperature inside the aviaries during the hot summer months in Sydney.

5. *Wire Mesh*

Heavy gauge wire mesh is essential to contain most cockatoos. Usually 14 gauge is adequate, although some individuals will even demolish this. The largest advisable size is 2.5 cm (1 in.) square as anything larger will allow sparrows to have access to aviaries. Bob Lynn prefers 2.5 cm (1 in.) chain wire, which is ideal but has become just too expensive in comparison to weldmesh. Bob does recall one escape artist who even chewed holes in this chain wire.

A recent trend in cockatoo aviary construction is to use 12 mm (½ in.) square by 16 gauge galvanised weldmesh. The theory is that the mesh is too small to allow a cockatoo to get its beak in to start chewing. I recently saw an aviary constructed of this mesh being systematically demolished by a pair of Red Tailed Black Cockatoos — there were patches on the patches on the patches. I feel the best available wire mesh for cockatoos and other large psittacines is 2.5 cm (1 in.) square (14 gauge) hot dipped galvanised weldmesh sheets. Note I stress hot dipped which means the wire mesh is galvanised after it is made. Do not use mesh which is made from galvanised wire as it rusts very quickly at the joints.

6. *Partitions*

When, as is usually the case, aviaries are constructed in groups or rows, consideration must be given as to how the aviaries are to be separated from each other. Breeding pairs of cockatoos can be aggressive towards their neighbours and are capable of inflicting serious injury to feet, legs, beaks and even wings through wire partitions. Therefore it is necessary to provide double wire partitions with a space between which is wide enough to prevent a bird's foot being bitten by its neighbour. Up to 10 cm (4 in.) may be needed for the larger cockatoos.

Solid partitions are my preferred alternative, which eliminate all chances of maiming as well as providing the privacy conducive to successful breeding.

7. *Floors*

Various floors are used for cockatoo aviaries each with their own advantages and disadvantages. Natural earth floors are probably the most commonly encountered floors in cockatoo aviaries. They can be attractive if the occupants allow some flora to grow, but the disadvantages are immense. Earth floors are difficult to clean and they are a constant media for reinfestation with intestinal worms. They provide ideal conditions for vermin such as rats and mice which in turn attract snakes in rural areas.

Sand-filled and pebble floors have a pleasant appearance but must be provided with suitable subsurface drainage to prevent them from becoming waterlogged during periods of rain. They are not easy to clean, particularly for cockatoos as their large droppings have a tendency to clog. In the long term there is usually a silting up effect which limits the absorbency and hence the effectiveness of this floor covering.

Concrete floors, which probably have the least pleasing aesthetic appearance, are without a doubt the most functional aviary flooring. They can be reasonably attractive and are easy to clean if kept covered with a sprinkling of sand. Concrete floors help to eliminate vermin and worm infestation as well as facilitating cleaning with a broom and shovel and/or by the use of a high-pressure water hose.

When I put the question to Bob Lynn of what type of floor he would use if building his aviaries again, He replied as follows: "I like the natural earth floors with grass growing where possible but the troubles I've had with rats and mice over the years have made me think twice about them.

"If I was building aviaries again I would start with a concrete raft foundation and floor, then build a brick wall on top of the floor around the perimeter of the aviary to a height of approximately 50 cm (20 in.). The aviary framework would then be built on top of this wall. I would then fill about 30 cm (12 in.) of soil over the concrete floor and plant various grasses. This type of construction would prevent rats from tunnelling into the aviary from outside and make it easy to flood mouse holes."

8. *Aspect and Protection*

Aviaries in the southern hemisphere should ideally face north or at least in the case of large broad cockatoo aviaries have one side facing north so as to provide maximum sunlight.

Full protection from all prevailing winds is adviseable as well as fully enclosed weatherproof shelters, open only at the front and situated on the southern end of aviaries facing north. On aviaries facing east or west there should be fully enclosed shelters on both ends. Roofed areas should be provided at both ends of the aviaries to afford maximum protection to front and rear perches from weather conditions and predators.

Aviaries should never be built to face south and must always be totally protected on the southern side.

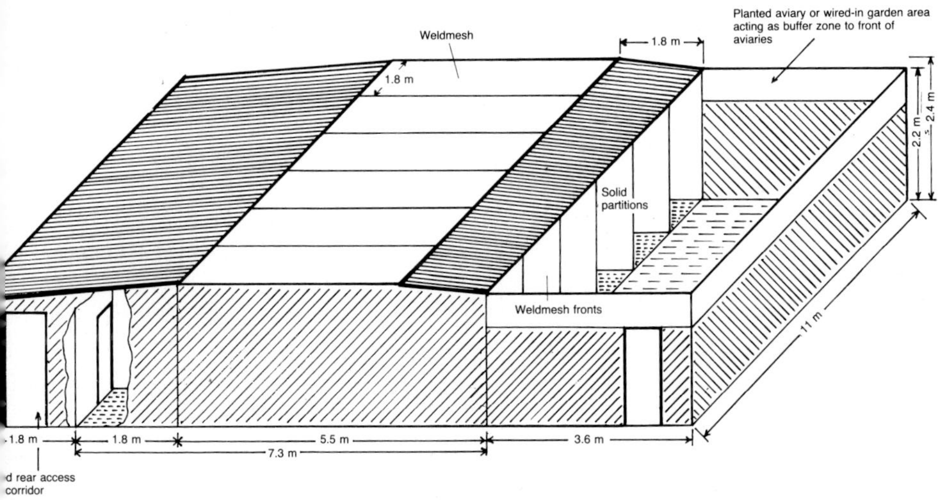

Figure 1. An ideal cockatoo breeding aviary complex.

9. *Walkways and Service Corridors*

Ideally all aviary complexes should be provided with front walkways or fully roofed rear service corridors or both. Walkways or service areas situated at the front of an aviary complex faciltitates feeding and cleaning and eliminates the possibility of escapes as well as preventing predator attacks and general disturbances to the front of the aviaries.

Fully roofed rear service corridors provide even easier cleaning and feeding facilities under all weather conditions. They also allow the use of external seed hoppers, watering devices and nesting logs or boxes for easy access direct from the service corridor.

For comfortable working conditions front service areas should be a minimum of 1.2 m (4 ft) wide. Rear service areas which provide external feeding and nesting facilities should be a minimum of 1.8 m (6 ft) wide.

10. *Adviseable Minimum Aviary Sizes*

The minimum aviary dimensions suitable for small cockatoos from Sulphur crested size down is 5.4 m (18 ft) long, 0.9 m (3 ft) wide and 2.15 m (7 ft) high.

Black Cockatoos (Calyptorhynchus) are best housed in the minimum dimensions of 7.2 m (24 ft) long , 1.8 m (6 ft) wide and 2.4 m (8 ft) high.

Palm Cockatoos' aviaries should be a minimum of 9.2 m (30 ft) long, 2.4 m (8 ft) wide and 2.7 m (9 ft) high.

11. *Feeding and Watering*

Aviaries should be designed to provide feeding or watering areas that are not subject to pollution or contamination by the birds' droppings. They should also be protected from rain, direct sunlight and vermin. These requirements are often achieved in ingenious ways by enterprising aviculturists. Basically there should be no possibility of birds perching over feed trays or water containers. The containers must be under shelter and the feed containers suspended in such a way as to prevent access by mice or rats.

12. *Snake Proofing*

Most aviculturists who live in rural areas of Australia have encountered the problem of a snake in their aviaries. Although I have had almost every conceivable disaster occur in my aviaries, I have never had a snake problem. This is one problem at least that city dwellers are free of, hence my knowledge in this area is all second-hand. Obviously fledged and adult cockatoos are not vulnerable to snake attack except for the unlikely possibility of being bitten by a venomous species. It is the eggs and nestlings that fall easy prey to snakes when they gain entry to an aviary, just as they do in the wild.

Everyone that I have discussed this problem with agrees that it is incredibly difficult to exclude snakes from an aviary. They can squeeze through the tiniest hole and can scale vertical walls if there is the slightest protrusion to aid their ascent.

Smooth surfaced sheeting such as flat sheet metal, fitted around the base of an aviary to the height of of 1.2 m (4 ft) will help to deter them.

Electric cattle fences have been used successfully in Queensland as a snake deterrent. The electrified wire is fitted close enough to the side of the aviary to make it impossible for the snake to pass inside it without making contact. Likewise if it goes outside the wire it will also make contact.

Suspended Aviaries

A suspended aviary is one that is constructed of weldmesh and supported at a suitable height above the ground to facilitate cleaning underneath. The concept is a very old one which was employed by the ancient Chinese with their split bamboo cages. In more recent years this type of housing has become popular for Psittacines in Europe and the U.S.A. I use them almost exclusively for housing lorikeets.

Suspended aviaries are obviously best located in bird-proof enclosures, inside large aviaries or with service doors opening into secured bird-proof service areas, for ease of feeding, etc., and to eliminate the possibilty of escape.

Currently in Australia some aviculturists are using this type of housing for Cockatoos with mixed success. Jim Gill has been successful with Sulphur-crested Cockatoos, Short-billed Corellas and Western Long-billed Corellas. The birds are housed in individual suspended aviaries which are just 3 m (10 ft) long, 1.2 m (4 ft) wide and 1.2 m (4 ft) high. They are made totally of 25 mm (1 in.) square by 12 gauge galvanised weldmesh sheets. They are situated in a secured area predominately under shelter with the front section exposed to the weather and with the nesting logs fitted externally. A number of pairs of Eastern Long-billed Corellas housed under the same conditions have so far refused to breed.

Matthew Crane of Sydney has been successful in breeding Gang Gangs housed in a similar sized suspended aviary situated in a totally sheltered area.

I am not convinced that suspended aviaries provide the ideal environment for cockatoos, although breeding successes are becoming more frequent. The advantages and disadvantages for this type of housing are as follows.

Advantages

1. Provides an easy to clean, hygienic environment.

2. Eliminates the occupants access to their droppings and stale food items, all of which fall through the bottom of the aviary.
3. The external nesting sites facilitate easy inspection and cleaning without disturbance to the birds.
4. Worming is not necessary when birds are housed in suspended aviaries.
5. The cost of erecting suspended aviaries is far less than for conventional aviaries.
6. More birds can be housed in a limited space.

Disadvantages

1. My observations suggest that cockatoos housed in this system do not feel as secure as when housed in secluded conventional aviaries.
2. Catching of cockatoos housed in this system often requires two people.
3. Open mesh suspended aviaries provide no protection from draughts, cold winds or general cold conditions, particulary when situated outside enclosed areas.
4. Efficient automatic watering systems are difficult to organise in suspended aviaries.

Battery systems of suspended aviaries, as I use for Lorikeets, are not practicable for cockatoos due to fighting and maiming through the wire partitions. The use of solid sheet metal partitions to provide privacy and eliminate fighting with neighbours deserves some investigation and experimentation.

Summary

Although I am an advocate for the use of suspended aviaries for many species and am particularly impressed with the conditions they provide, I feel that most cockatoos require more space, privacy and security for successful propagation than this type of housing can provide. Very few of the breeding successes that have occurred in the suspended aviary system have been total successes. In most cases the young have had to be taken for hand rearing at some stage during the rearing period. This suggests to me that the parents did not feel totally secure in their environment.

Housing for Cockatiels

Housing for this, the second most prolific of the Australian psittacines is by no means critical. In fact, breeding successes have occurred in every conceivable situation and often under the most adverse conditions.

Many years ago when I was involved in helping to establish the Lutino mutation in Australia, I found that pairs housed individually in small aviaries about 2.15 m (7 ft) long, 90 cm (3 ft) wide and 2.15 m (7 ft) high, provided me with good controlled breeding results. These small aviaries were fully roofed, had concrete floors and solid partitions between them.

Small suspended aviaries or even cages have also been used with good results. They can be situated indoors, under semi-shelter or even in exposed conditions with reasonable shelter fitted to the cage.

Aviary design and construction materials are not as critical with cockatiels, provided reasonable shelter is available and sanitary conditions can be maintained. These birds are hardy and their chewing abilities will inflict only minor damage to aviaries of timber construction.

DIET

Providing a nutritious, well balanced diet for cockatoos is a comparatively simple matter, but enticing them to eat it is quite often an impossible task.

The food a particular species, pair or individual likes the most is more often than not unsuitable for their long-term well-being. In many cases this particular food may be the only item they will eat. The common example is the addiction many cockatoos have for sunflower seed.

Of course it is impossible for me to suggest a dietary combination that will be accepted by all cockatoos and yet still provide a balanced diet suitable for each species. The food sources in the wild obviously vary immensely where so many distinctive species inhabit such diverse climatic regions.

On a number of occasions, usually after the stimulation of an overseas trip, I have attempted to develop a super nutritious, vitamin packed, mash diet which could be used as a supplement to the cockatoos basic diet, similar to the diets developed for lorikeets. Alas, each newly developed mash has been abandoned after months of perseverance because my cockatoos have refused to eat it. So I can offer no sensational new diet, only basic dry seed mixtures suitable for individual species and groups. In addition to this I feed an array of supplementary foods that can be offered and when an item is accepted, discretion as to the quantity fed in relation to its nutritional value must be used.

My own tendency is to feed a domesticated diet which is constantly available the whole year round and not rely heavily on the gathering of seasonal items such as ripening seed heads, berries, nuts, fruits, etc. These are used as treats, not essential ingredients to the diet. Of course, people living in rural areas have greater access to such supplements and may tend to rely more heavily on their use, particularly during the breeding season.

Bob Lynn relies heavily on the gathering of native seed cones and pods from Banksia, Casuarina and Hakea, which are available on the outskirts of Sydney all year round, as a supplementary food for his Black Cockatoos.

Supplementary Feeding

Ideally the nutritional supplementary foods should form a major part of the cockatoo's diet and the basic dry seeds should be a secondary food source; unfortunately in most instances this is not the case. Perseverance with feeding any or all of the supplementary food items suggested in the following list must be advantageous in attempting to supply the necessary dietary needs.

Suggested Supplementary Food Items

Raw peanuts, various other nuts such as walnuts, almonds, brazil nuts etc., sprouted seeds such as sunflower, corn, millets, sorghum, etc., sweet corn, wholemeal bread, dog kibble (soaked or dry); fruits such as apple, pear, orange, banana; vegetables such as silverbeet, broccoli, cauliflower, green peas, beans, carrots, fresh pumpkin and melon seeds; table scrap bones such as chop bones, chicken bones, lamb, pork or beef bones etc., retaining as much meat on them as possible; native and introduced seed pods and cones, fruits, berries and the various seeding grasses.

Species Diets

The following suggested diets for individual species and for groups are maintenance diets only, that are to be used during the majority of the year to maintain the birds in a healthy lean condition. As the breeding season approaches the supplementary food items most preferred should be increased, then fed on an unlimited basis when young are being reared.

Palm Cockatoo

A basic dry seed diet of predominately sunflower seed together with a little safflower and maize.

Suggested supplementary foods are: raw peanuts, walnuts, almonds, brazil nuts, sprouted sunflower and maize, dry kibble, sweet corn, silverbeet, green peas and table scrap bones. Bob Lynn also fed his Palm Cockatoos a wide range of native seed bearing cones and pods such as Banksia, Casuarina, Hakea, Native Pear, etc., of which *Hakea servica* was the most favoured.

Bob actually saw his Palm Cockatoos kill and eat a rat that had managed to gain entry to their aviary. I am not suggesting that Palms should be fed live rats, but obviously animal protein is appreciated at times.

Black Cockatoos: Calyptorhynchus

A basic dry seed diet of sunflower seed together with some safflower seed.

Supplemented daily with raw peanuts, sprouted sunflower seed, sprouted maize, green peas, apple and silverbeet to which should be added any other of the supplementary foods that may be accepted, on an irregular basis.

Native seed bearing cones and pods should be supplied on a daily basis, not only for their nutritional value but for the therapy provided by the chewing.

Each species has its preferred variety. Red-tails prefer Casuarina cones (of the Casuarinas growing in the Sydney region *Casuarina littoralis* are liked the most) and will sometimes chew up Banksia cones and Hakea pods. As few as six Casuarina cones per bird each day will satisfy their needs.

Glossy Black Cockatoos are obsessed with Casuarina cones and in Bob's opinion prefer *Casuarina littoralis*, above all others growing around Sydney. Six cones per bird each day is sufficient for their needs although they will devour far greater quantities if given the opportunity. Little attention is paid to other native food by this cockatoo.

Yellow-tailed Black Cockatoos are less specialised feeders and will readily accept Banksia cones (*Banksia serrata* and *Banksia ericifolia* are the most preferred from the Sydney area), Casuarina cones (*Casuarina littoralis* as preferred by other Black Cockatoos), Hakea *(Hakea servica)* and cones from various species of introduced pines. I feed my Yellow-tails a few Casuarina cones and two Banksia cones per pair daily as well as Hakea and pine cones on an irregular basis and have found this adequate to maintain them in superb condition.

White-tailed Black Cockatoos are usually not interested in Casuarina cones, preferring Banksia cones *(Banksia ericifolia)* and Hakea seeds, as well as various pine cones. Two Banksia cones daily, together with Hakea fruit and pine cones fed every few days, are adequate for one pair.

Black Cockatoos are not usually prone to obesity so it is not necessary to limit fattening or oily foods such as sunflower seed.

Gang Gangs

This cockatoo has probably caused more heartache and concern to its owners than any other species, due to the habit of feather plucking, for which unfortunately Gang Gangs have become renowned. Many expert aviculturists have related this habit to dietary problems and to the feeding of sunflower seed in particular — I cannot agree.

During my 30 years of keeping Gang Gangs (who by the way are my favourite bird), I have had many instances of feather plucking, none of which I can relate to diet — boredom, discontentment, emotional trauma, partner incompatability, and many other psychological problems, yes, but in my opinion seldom, if ever, dietary.

Many aviculturists will not feed sunflower seed to Gang Gangs. I feed mine a basic dry seed diet of sunflower seeds together with sprouted corn, raw peanuts, green peas, wholemeal bread, apples, silverbeet and a few Casuarina cones, as well as all the table scrap bones available.

Any of the other food items on the supplementary list must be advantageous if accepted. Banksia, Hakea and Acacia, as well as many other native seeds, pods and cones, are often relished and are also valuable for the therapy they provide in chewing material. The introduced Hawthorn, Cottoneaster and Pyracantha berries are all appreciated when in season.

For those who would rather not feed their Gang Gangs sunflower seed I would suggest a mixture of 50% canary seed, 20% safflower, 10% French white millet, 10% oats and 10% corn.

Galah

This cockatoo is subject to obesity after reaching maturity, particularly in warmer climates. I have found that all the oily and fattening seeds, such as sunflower seed, canary seed, oats, etc., must be totally eliminated from their diet if they are to be maintained in a lean and healthy condition. Even then it is usually necessary to limit the combined dry and sprouted seed fed daily to one large handful per bird. My suggested dry seed mixture for this species is equal parts of French white millet, Hungarian millet (pannicum) and Japanese millet, although I now only use the larger French white millet as it is usually the most preferred.

The supplementary foods should consist of sprouted French white millet, sprouted corn, green foods such as silverbeet, broccoli, cauliflower, seeding grasses, etc., and one pod of green peas daily. I have found that reasonable quantities of green food can be fed without adversely affecting the Galahs' weight, although monitoring is advisable.

As the breeding season approaches the limited seed diet for breeding pairs should be gradually increased to double that of the maintenance diet to help promote breeding condition. Of course, when young are being reared an unrestricted diet should be fed which includes dry and sprouted sunflower seed.

Galahs kept in cool climates may not require such severe dietary restrictions.

Sulphur-crests, Major Mitchells and all Corella Species

I have found the dietary requirements of these species to be similar and have never encountered an obesity problem in this group when their housing provided adequate room for exercise.

Most avicultural writers suggest a basic mixed seed diet of maize, wheat, oats, milo, etc., with little or no sunflower seed for this cockatoo.

This type of seed diet is exactly what most people who know me and my approach to diets, would expect me to suggest for a basic dry seed diet, as it is low in protien, non oily and totally uninteresting, just as my basic seed diets are. I am sure it will maintain this group of cockatoos in good condition, providing adequate supplements are supplied. But, just to be perverse, I use a basic dry seed diet of sunflower seeds for these species of cockatoos.

Recently I conducted an experiment over a 12-month period where two pairs of Major Mitchells were housed in identical aviaries, which provided adequate exercise. One pair was fed a basic diet of sunflower seed and the other, French white millet. Both pairs were fed identical supplementary foods although their preferences for these foods varied considerably.

During the period of this experiment neither pair nested, although both pairs worked their log.

After the 12-month period both pairs were caught up and examined closely. I could detect no marked variation in weight, feathering or general condition, which leads me to believe that the all-important factor in diets of cockatoo species, which are not subject to obesity, is the quality and quantity of the supplementary foods supplied, not the basic dry seed diet.

My suggested supplementary food items for this group are:- sprouted sunflower seed (one handful per pair per day) sprouted corn in whatever quantities accepted, about 12 raw peanuts per day, two or three pods of green peas per day, silverbeet, 1/4 slice of wholemeal bread per pair per day and a slice of apple, together with sweet corn, broccoli, cauliflower, carrot, dog kibble, pumpkin and melon seeds, table scrap bones and seeding grasses, all when available and when accepted.

Cockatiels

Although the Cockatiel falls broadly into the cockatoo family, its dietary requirements are similar to those of the Australian grass parakeets. Like the grass parakeets they will become overweight if fed on oily seeds such as sunflower and canary seed. Either a dry seed diet of equal parts French white millet, Hungarian millet and Japanese millet, or as I prefer, just French white millet is recommended.

The suggested supplementary food items for Cockatiels on a maintenance diet are: sprouted sunflower seed, limited to a dessertspoonful per pair per day; sprouted millet; 1 pod of green peas per pair per day; silverbeet; cauliflower and broccoli, if accepted, as well as apple and seeding grasses.

These supplementary foods should be increased as pairs come into breeding condition and fed in unlimited quantities when rearing young.

Storage of Seed

Seed should be stored in vermin proof, air-tight containers such as specially constructed storage bins or metal or plastic drums with secure lids.

Storage of Supplementary Foods

The storage life of supplementary foods such as vegetables, green foods, fruit, bread, etc., is obviously extended considerably if kept under refrigeration. Rapid growth of sprouted seeds, during periods of hot weather can be inhibited and souring prevented by refrigeration.

Nuts are best kept in an airtight container and stored in a cool, dry place.

Banksia cones and pine cones may be stored for months at a time in mesh bags kept in a cool, dry place.

Casuarina cones may be kept for weeks at a time, particularly in cool weather, by placing branches of cones in tubs of water situated in shaded areas. Bob Lynn has found that if the Casuarina cones stored in this way start to open and expel their seeds, during periods of heat, the process can be halted and reversed if detected early, by immersing the seed cones themselves into the tubs of water. The opening cones treated in this manner will close again, retain their seed and have their storage life extended for a week or two. Bob has found that Casuarina cones stored in this manner are preferred by the Black Cockatoos.

I have found that freshly picked Casuarina cones will last for some months if immersed in a bucket of water for 12 hours, drained off and then stored in open containers or mesh bags under refrigeration. The Yellow-tailed, Glossy and Red-tailed Black Cockatoos as well as Gang Gangs in my aviaries, all accept Casuarina cones stored in this manner.

Hakea seed pods may also be kept by placing the branches of pods in tubs of water, similar to Casuarinas, and will also keep for long periods of time under refrigeration.

Sprouting Seed

Sprouted seeds are a popular and, in my opinion, an important method of providing added nutrition as well as many essential vitamins such as niacin, riboflavin etc., to the diet of average birds. For once a seed sprouts it again becomes a living thing and its entire chemical composition changes.

Most seeds may be sprouted by soaking for 24 hours in water to which is added an anti bacterial-fungal agent such as "Milton", a product used for the sterilisation of babies' bottles. Some aviculturists use bleach for this purpose but Jim Gill feels that the lack of quality and strength control imposed on household bleaches renders their use undesirable.

The soaked seed is then tipped into a perforated plastic icecream container or colander, thoroughly washed, then allowed to sprout in this container. The duration of the sprouting period varies according to the variety of seed and with weather conditions. Sprouting during the winter will often take one or two days longer than in the summer, but may be hastened by placing a lid or sheet of glass over the container of seed.

There is no advantage in allowing sprouted seed to grow long shoots, for the maximum nutritional value is reached as soon as the shoot breaks out of the seed and because most birds are only interested in eating the seed kernel, not the shoot.

Hand Rearing Diets

There are hundreds of hand rearing diets in use around the world. Most are adequate, in my opinion, although often excessive in oils and protein. Some are simple while others are incredibly complicated and time consuming to prepare.

As a practical, or is it lazy, aviculturist, my requirements for a hand feeding diet are:

1. It is simple to prepare.

2. Has no ingredients that will deteriorate or promote bacterial or fungal growth while being stored.
3. Requires no cooking or storing under refrigeration, in other words it is a dry mixture.
4. The diet provides all the nutrition required to rear healthy chicks.
5. The protein content of the hand rearing mix is approximately 15-20%, which laboratory tests in the U.S. suggest is optimum for maximum growth rate in young psittacines.

Hand Rearing Diet

1 part chicken starter crumbles (approximately 20% protein).

1 part egg and biscuit canary rearing food.

1 part sunflower meal.

1 part Farex.

A multi-vitamin and mineral powder such as "Ultrameal" or "Ornithon" should be added to the mix in the ratio of 1 level teaspoon to 4 cups of the mixture.

This dry mix is best kept in an airtight container and stored in a cool, dry place.

To prepare ready to feed just add hot water and allow to cool to a suitable temperature. Add a calcium supplement such as Calcium Sandoz or D.C.P. 340 powder to one feed per day and a liquid vitamin supplement such as Penta-vite or Avi-drops to another.

Hand Rearing Diet for Day Old Chicks

One part strained mixed vegetable baby food (Gerber or Heinz), one part strained apple baby food (Gerber or Heinz) mixed together and diluted with crop wash (see Hand Rearing section of Management chapter) and fed lukewarm for the first three feeds.

After the first three feeds and providing the chick is digesting well, one part Heinz rice cereal should be added to the mixture and the crop wash discontinued.

If the chick is still digesting well after three more feeds, replace the rice cereal with Farex.

On the third day, providing digestion is still complete between feeds, the Farex should be gradually replaced over a few feeds with a mixture of one part ground chicken starter crumbles (approximately 20% protein), one part ground egg and biscuit canary rearing food and one part Farex. Also at this stage a multi-vitamin supplement (2 drops of Penta-vite) should be introduced into one feed per day and a calcium supplement (2 drops of Calcium Sandoz or a small pinch of D.C.P. 340 Calcium Powder) to another. Of course as the chick grows the supplements should be increased accordingly.

During the next day the apple should be reduced by half and the mixed vegetables discontinued over the next few feeds.

Providing the chick continues to do well, the remaining apple should be discontinued and the rearing mixture gradually increased to that of the normal hand rearing diet, except with the granular ingredients ground fine, so as to reach full strength from seven to 10 days old. Grinding of the ingredients is usually unnecessary after two weeks old.

Bob Lynn's Hand Rearing Diet

Bob uses this diet to hand rear reasonably advanced baby Black Cockatoos, or as supplementary food for young Black Cockatoos whose parents neglect their feeding duties after the youngsters fledge.

This diet is not suitable for chicks under seven days old.

1 cup ground sunflower kernels
¼ cup ground raw peanuts
¼ cup ground almonds
¼ cup wheat germ
¼ cup ground canary seed
4 crushed wheatmeal biscuits
1 tablespoon glucose powder
1 tablespoon calcium powder
1 tablespoon cornflour

All mixed together and stored in a dry, airtight container. To feed, mix with hot water to the consistency of a wet porridge.

Hand Rearing Diet For Glossy Black Cockatoo

Bob has found his usual hand rearing diet unsuitable for Glossy Black Cockatoo chicks, due to the specialised feeding habits of this species and suggests the following formula for this purpose.

1 tablespoon sunflower meal
1 tablespoon raw peanut meal (peanuts must be dehusked before grinding)
2 teaspoons almond meal
2 teaspoons Farex
2 teaspoons egg and biscuit canary rearing food

It is essential that all ingredients are finely ground and sieved, as young Glossys have difficulty in swallowing and digesting coarse feed. All ingredients are mixed together dry and made into a porridge with warm water. A calcium supplement and multi-vitamin and mineral supplement are added to one feed each day.

MANAGEMENT

Cockatoos are a relatively hardy and robust group, that when provided with clean and adequate housing, a good diet and a reasonable management routine, should present comparatively few problems.

Cleanliness is an important factor with all captive livestock and cockatoos are no exception. Aviary floors should be cleaned weekly, particularly under perches. Floors, walls, perches, wire mesh as well as food and water containers, should be scrubbed regularly with an iodine based disinfectant such as "Sani Chick".

Floor cleaning may consist of dry cleaning carried out by scraping, sweeping and periodic liming of the surface, or, if the floors are well drained, high pressure water hoses may be well employed.

Perches must be positioned to provide the maximum flying distance available, as well as to avoid fouling of food and water containers. Regular replacement of perches is often necessary for many species of cockatoos, due to their incredible chewing abilities. Even robust, well seasoned Australian hardwood perches are often reduced to splinters in a few weeks, but the temptation to replace wooden perches with a hard composite or steel pipe perch must never be considered, as chewing is good therapy and hard or metal perches can seriously affect a birds feet.

Water Containers and Facilities

Although many species of Australian cockatoos will not bathe it is advisable to provide a water container of adequate weight and capacity to prevent it from being tipped over and emptied by active cockatoos.

Bob Lynn always provides large water containers at least 60 cm (2 ft) in diameter and 10 cm (4 in.) deep for his Black Cockatoos although the only species he has observed bathing is the White-tailed and Yellow-tailed Black Cockatoos.

Containers should have a smooth, non-porous, glazed surface to hinder the growth of algae and to facilitate cleaning.

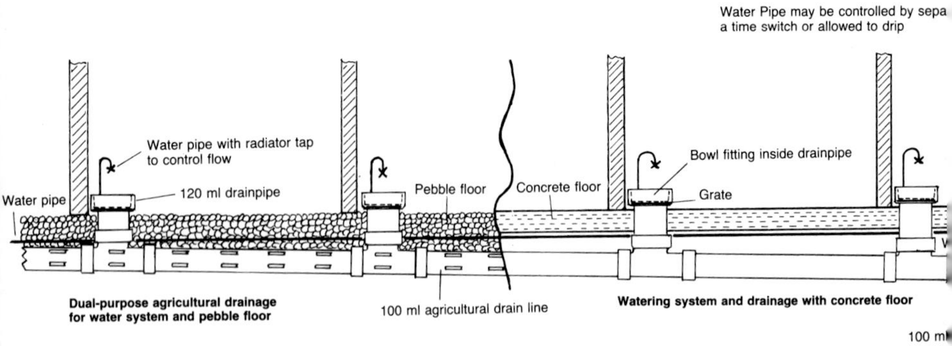

Figure 2. Watering system.

When automatic watering systems are used, suitable and adequate drainage must be provided to prevent constantly wet floor areas. (See Figure 2.)

A year or two ago when visiting the Gosford area of New South Wales to speak at the Central Coast Avicultural Society Meeting I was treated to an inspection of a prominent member's extensive parrot collection. He pointed out that he used the very efficient W.W.W. watering system. When I inquired what the W.W.W. stood for he laughingly replied "Wife With Watering-can".

Feeding Facilities

I favour the use of a hopper fitted with a large tray which serves the dual purpose of catching the seed husks as well as providing an area to feed the supplementary foods (see Figure 3). These large hoppers and trays are constructed of sheet galvanised iron and provide an easy to clean, labour saving device, which I have found to work well for all species of cockatoos.

When seed is provided in dishes, they should be placed in large suspended metal trays, such as inverted metal garbage bin lids, which catch the husk and scattered seeds as well as catering for the supplementary foods.

All feeding facilities must be situated in a totally dry position which also excludes access to rats and mice.

Regular emptying and cleaning of trays and all feeding utensils is essential, as stale food and dusty or mouldy seed husk can promote fungal and bacterial infections.

Floor Surfaces

Although floor surfaces have been discussed previously, the necessity of clean, dry aviary floors cannot be overstressed.

Concrete floors sprinkled with sand provide a reasonably easy to clean surface, as the sand prevents droppings, etc., from sticking to the floor. Well drained concrete floors are easily cleaned by regular hosing.

Pebble or lucky stone floors, when provided with a good below surface drainage system, are also easily cleaned by regular water hosing.

Sand filled floors require a suitable sub-surface drainage system to keep them dry (see Figure 2) and may be cleaned by removing the soiled surface regularly. Total replacement of the sand filled floors is usually necessary every year or two.

Earth floors require constant attention, often becoming infested with vermin such as rats and mice. Regular raking and liming of the soil is advisable, particularly for the ground frequenting species of cockatoos.

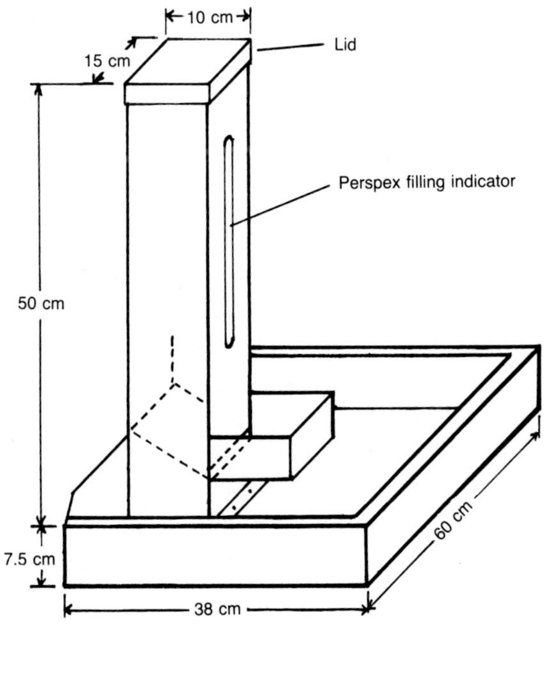

Figure 3. Food tray and seed hopper.

Worming

Regular worming of all cockatoos housed in conventional aviaries is essential in my opinion, although Bob Lynn only worms his Black Cockatoos occasionally, if he suspects there is a worm problem. Perhaps this suggests that Black Cockatoos, who spend little or no time on the ground, are not as susceptible to worm infection as other cockatoos.

I have seen worm infections in Galahs, Sulphur-crests, Major Mitchells, the Corellas and Cockatiels, all of which spend much of their time on the aviary floor.

Cockatoos housed in suspended wire aviaries should never need to be wormed, whereas those housed on earth floors should preferably be wormed every three months, and those housed on all other floor surfaces every six months. It must be considered good practice to regularly worm all species, regardless of their susceptibility.

The only practical and effective way of worming cockatoos is to catch up each individual and administer the worming agent direct to the crop, using a metal crop needle attached to a syringe. Worming via the drinking water is totally ineffective, as cockatoos are capable of drinking little or no water for a number of days if the water is deemed unpalatable.

I have found the most effective worming agent for cockatoos to be "Panacur 2.5" which is a sheep drench marketed in Australia by Hoechst. The recommended dose is 1 ml for the small cockatoos such as Gang Gangs, Galahs and Corellas, etc., and 1.5 ml for Black Cockatoos and Palm Cockatoos. Panacur is not registered as a worming agent for birds so the responsibility for its use must be accepted by the aviculturist concerned.

Panacur is not toxic, and has a large dose rate tolerance which provides a safe working margin. The occasional death that occurs after worming with Panacur is due to an intestinal blockage caused by clusters of worms from a massive infestation. Infestation of this magnitude cannot be treated successfully. I make a practice never to worm a bird on a hot day as I once lost two Australian Parrakeets from stress after worming on a day over 30°C.

Routine Daily Observation

The practice of observing all your cockatoos at least once and preferably twice a day should become routine. Any inactive, dull and listless bird should be caught up and examined for loss of weight, dirty vent, canker (fungal) infections in the mouth, beak abnormalities or injuries and the appropriate measures taken. Prompt diagnosis and treatment is essential if a satisfactory recovery is to be achieved.

Any bird requiring treatment or observation is best placed in a cabinet, holding cage, or small aviary preferably in a bird room or garage or some other enclosed warm place, to simplify medication and further surveillance.

Nest Facilities and their Cleaning

In Australia the nest sites provided by most aviculturists for their cockatoos are natural hollow logs of eucalypt, or other similar native hardwoods, which are resistant to the destructive chewing abilities of our cockatoos. (See Plate 1.) The logs should be adapted to provide access for inspection and cleaning purposes through strategically placed doorways or removeable lids. Such accesses can be cunningly constructed by cutting out a complete section of the log and fitting it with hinges and a lock so as to form an almost undetectable door, or by cutting a suitable hole and providing it with a metal door.

However inspection facilities are provided, it is essential that all nesting sites have accessibility to the nest chamber so that any servicing required during the breeding period can be carried out with minimal disturbance.

When hardwood hollow logs are unavailable, alternate nesting sites must be provided in the form of barrels or nest boxes, sturdily constructed of the hardest timber available, and of dimensions similar to those suggested for nesting logs in the Species chapters. For the larger species of cockatoos it may be necessary to reinforce and protect vulnerable chewing areas with strips of sheet metal.

Metal garbage bins are commonly used for nest sites overseas and are obviously a simple and durable solution to the problem, providing the cockatoos accept them.

Nests may be filled with a variety of materials, such as wood dirt taken from the inside of rotting trees; crushed soft termite mound; fine wood shavings; peatmoss etc.

The cleaning of cockatoo nests is usually unnecessary during the rearing period as they are clean, dry, nesters. Prior to the breeding season all nests should have at least the top 10 cm (4 in.) of the nest filling replaced, be washed down inside and out with a disinfectant, and when dry sprayed with a mite repellant such as "Duramitex" or some Maldison based product

A △

B ▷

D ▷

C ▽

E △

F △

Plate 1: Nesting Logs.

A First totally parent reared Palm Cockatoo nesting log. Note chewed sticks deposited in log by breeding pair.

B Red-tailed Black Cockatoo nestling in its standing, vertical hollow log.

C Female Glossy Black Cockatoo at natural entrance spout of vertical breeding log.

D Breeding pair of Gang Gangs at their hanging, vertical nesting log.

E Galah at its hanging, vertical nesting log.

F A breeding pair of Major Mitchells at the entrance spout of their horizontal nesting log.

such as "Malathon", which is a garden spray produced by Chemspray Pty Ltd for the control of aphids, thrips, caterpillars, etc., and when diluted with 50 parts water to 1 part Malathon is an effective non-toxic anti-mite spray that remains active on the nest surfaces for up to 12 months.

Pairing

As most species of Australian Cockatoos can be readily sexed, obtaining a true pair does not present the difficulties encountered in some psittacines, except of course for the Corella group and those Sulphur-crested females whose eye colour does not change. Surgical sexing is advised for these species as much time can be wasted trying to establish a pair.

Compatibility of pairs is of the upmost importance in most species of cockatoos, if successful breeding is to be achieved. Close scrutiny of unestablished pairs is necessary throughout the year to detect any signs of incompatibility and changes made when needed.

Egg Breaking

Egg breaking is not an uncommon occurrence with nesting cockatoos which can sometimes be controlled by providing a deeper or partially inclined nest site.

Some years ago I had a male, dilute White Galah who insisted on breaking each egg soon after the female laid it in their hollow log which was about 60 cm (2 ft) deep. The following season a 1.5 m (4 ft 10 in.) long, vertical log with an inspection door near the bottom was provided — the eggs were not broken and the young were reared.

If the eggs are broken and eaten the bird concerned has a vice which will probably never be cured or even improved, although I have been told of cures which involve the replacement of fresh eggs with rotten addled eggs or with dried out eggs that have been filled with mustard from a syringe. I cannot personally vouch for this cure.

Rearing Chicks

As soon as the chicks hatch I carry out daily inspections of the nest — twice daily when time permits. Many aviculturists do not agree with this constant interference, but my personal experience indicates that the advantages of constant surveillance far outweigh any disadvantages. From these constant inspections, potential problems are often rectified before they become disasters. For example, a chick that slips behind its nest mates may be taken for fostering or hand rearing, and youngsters being poorly reared by their parents, or suffering from vitamin deficiencies, are detected and taken for hand rearing before they reach the point of no return.

A suitable and adequate diet must be provided to pairs rearing young, and care taken to see that sufficient nutritious food items are being fed to the chicks. If in doubt provide more variation in the diet until rearing foods more to the parents' liking are found.

Fostering

Fostering of deserted chicks or chicks being poorly reared is not a common practice in Australia, probably due to the majority of cockatoos being large birds and easy to hand rear from an early age, this is the option taken.

When we are confronted with deserted chicks under four or five days old, or with fertile eggs that have been abandoned, fostering, if only for a few days, is the best way to go.

Of course, the employment of the same or similar species of cockatoos for foster parents is ideal, but I have found that a simple solution seldom occurs in aviculture and we are usually forced to be inovative in our choice of pairs to become foster parents. Don't despair if the psittacines available are quite dissimilar, providing they are large enough to cover the eggs or the young, there is a good chance they will incubate the eggs and or feed the babies at least for a few days.

Regular inspection of young being fostered is absolutely essential so that the first signs of discontentment or desertion by the surrogate parents may be detected and the young removed for hand rearing.

Hand Rearing

The ideal time to take chicks for hand rearing, if you have a choice, is at the early pin feather stage. Chicks at this age adapt readily to their surroundings and diet, require little artificial heat, and need only three or four feeds a day. Always avoid taking day-old chicks for hand rearing whenever possible, as chicks fed by their parents or foster parents for three or four days have a far better start in life.

When freshly hatched chicks have to be reared by hand I would suggest the following procedures.

The newly hatched chick must be placed in some form of brooder with an accurate temperature control and maintained at 97°F. If the brooder has no facilities for humidity control, a small container of water should be placed in the brooder to provide humidity and thus prevent dehydration of the chick.

Next the bacteria necessary for the satisfactory operation of the hatchling's digestive system must be introduced into the crop. This is accomplished by selecting one of the parents, or any other healthy cockatoo (not necessarily of the same species) and depositing several mls of warm, boiled water into the crop via a metal crop needle and syringe. The crop should be gently massaged externally to mix its contents with the water, allowed to settle, then as much of the resulting fluid as possible withdrawn with the crop needle and syringe.

When sucking the fluid from the crop, care must be taken to place the end of the needle in such a position in the crop so as not to block the end of the needle with large pieces of half digested food or with the wall of the crop. This process has generally become known as crop washing and the fluid known as "crop wash". Crop wash should be stored in a sealed container under refrigeration, and should not be kept more than three days.

This procedure may seem to be a daunting task to some, but after a few attempts it is possible to become reasonably proficient. I recently, through necessity, successfully crop washed a Macaw single handed.

Within two or three hours of hatching, a few drops of the lukewarm crop wash, the exact quantity depending on the species of cockatoo, but enough to slightly expend the empty crop, should be fed to the baby from a small spoon, the open end of a syringe, or any other feeding device which may be preferred. My preference is a small teaspoon with the sides folded up to form a chute.

Two hours later a small amount of food (as suggested in the Hand Rearing section of the Diet chapter) together with a few more drops of crop wash should be fed to the chick.

The two-hourly feeds should be continued until about 10 p.m., then one feed in the middle of the night and start again at 6 a.m. The two-hourly feed procedure with one feed in the middle of the night should continue for the second and third day, providing the crop is totally emptying between feeds. If the crop is not emptying fully, the quantity fed should be reduced, or the period between feeds be extended.

After the third night the "middle of the night" feed may be discontinued and the temperature of the brooder reduced by 1°F per day until 90°F is reached at about 10 days old. During the same period the food strength is gradually increased as suggested in the Diet chapter. Any step that appears to have an adverse effect on the chick should be immediately stopped, and the rearing process returned to the previous set of circumstances until the baby stabilises, before advancing again.

As the chick develops, and the crop capacity increases, the number of feeds required each day decreases. Usually by early pin feather stage (14 to 21 days old) five to six feeds a day are needed; at half feathered, three to four feeds a day and when fully feathered two or three feeds a day.

When the chick commences to pin feather, usually at about 14 days old, I prefer to remove it from the brooder and place it in a small cardboard carton, (approximately 25 cm × 25 cm × 30 cm (long)), which is provided with a heating pad (I use the heating pads from foot warmers marketed as "TV Booties") standing vertically at the side of the carton which is then lined with a few thicknesses of newspaper and paper towel. This allows the youngster to huddle close to the pad to keep warm or move away from it, if too hot. I cover the top of the carton with two thicknesses of towelling, leaving a small gap to allow air in and which can be adjusted to vary the temperature as required.

Chicks held under these conditions are subjected to small temperature variations, similar to those encountered by parent reared chicks. There is no need to provide water for added humidity when brooding chicks with this method.

Paper nest lining should be replaced every feed time and the chick must be thoroughly cleaned of all surplus food with toilet paper or similar soft disposable material after each feed.

Providing weather conditions are mild, and overnight temperatures are not dropping too low, heating pads can be dispensed with once the chick is half feathered as long as it is kept well covered to retain the heat generated by its own body within the carton.

Overnight the air slot should be reduced by the covering towel to about 10 mm. (½ in.) wide. Should any deterioration in the young be detected at this stage of the rearing, either in food digestion or general health, it is advisable to replace the heating pad immediately.

When the young is fully feathered it should be placed in a small indoor cage during the day and provided with a variety of foods (as suggested in the Hand Rearing section of the Diet chapter), then returned to its cardboard carton during the night.

The same number of feeds per day should be continued during this early weaning process, then as the chick starts to pick at the foods provided, the soft feeding should be reduced to one feed a day, given in the early evening to ensure the chick has a full crop to see it through the night. When the youngster is feeding itself and at least one heaped teaspoon full of food can be felt in the crop at early evening, the evening "soft food" meal should be discontinued, to force the chick to continue self feeding.

As soon as the youngster is independant it should be placed in a large indoor cage, preferably one metre or so long, for a couple of weeks before being placed in a small sheltered aviary: then some weeks later into a larger aviary more suitable for housing cockatoos.

Always remember when hand rearing cockatoos that all transition steps must be taken gently, and be prepared to retrace a step when all is not well.

Artificial Incubation

I would suggest that artificial incubation be resorted to, only when no other alternatives are available, as the hatching success rate of eggs artificially incubated for the full incubation period, can at times leave a little to be desired. No doubt experience and working knowledge of your own incubator increases the success rate.

If it is possible to foster the eggs under any suitable sized bird, not necessarily a psittacine (bantams or pigeons are often used successfully) for at least those critical first two weeks, before being placed in an incubator, the hatching results are greatly improved.

I have applied the following data and procedures to my Marsh Roll X Incubator with a reasonable success rate, and submit them as a guide.

Always thoroughly clean and disinfect, then operate the incubator for 24 hours prior to placing the eggs in it, so that any necessary adjustment to temperature and humidity can be made. The normal operating temperature for cockatoo eggs is 99°F, although some aviculturists prefer the slightly lower temperature of 98.5°F, with 59% relative humidity which is equal to a reading of 86°F on the wet bulb thermometer.

Before placing fresh eggs into the incubator always lightly clean and weigh each egg so that weight loss may be monitored during the incubation period. The generally accepted ideal weight loss of 16% over the whole period should be aimed for. This is calculated by dividing the initial weight of the egg by 6.25 to obtain the prefered 16% overall weight loss, which in turn should be divided by the number of days in the normal incubation period for the particular species, to ascertain the ideal daily weight loss. All of this information should be recorded for each egg prior to it being placed in the incubator.

The eggs should be placed on the incubator racks with the large, blunt end uppermost and inclined at approximately 45 degrees to the horizontal. If the incubator has no automatic turning device the eggs must be turned manually from a 45 degree inclination in one direction to 45 degrees in the other, commencing 24 hours after the start of incubation. A minimum of three turns per day (morning, midday and evening) is essential and additional turning advantageous, but the eggs must always be turned an odd number of turns each day so the eggs do not lay in the same position consecutively during the long night period. Automatic turners rotate the eggs regularly although I feel it is advantageous to manually turn the inclined eggs from one direction to the other twice daily in addition to the automatic turning.

After three days the eggs should be weighed again and their weight compared with the theoretically desired weight after three days loss. If the eggs have lost too little weight, this suggests the humidity is too high and should be reduced, whereas if the eggs have lost too much weight, the eggs are drying out too quickly, hence the humidity must be increased. This weight loss should be monitored closely and the weight checked every few days and the appropriate adjustments made to the humidity. Eggs often hatch successfully with considerably more or less weight loss than the advised 16% ideal.

Three days before the eggs are due to hatch, the humidity should be increased to register 92°F on the wet bulb thermometer (about 75% relative humidity), the turning of the eggs discontinued and the eggs placed in a fine wire or plastic mesh basket, to allow full air circulation around the eggs and to protect the chicks when they hatch. Some prefer to use a separate incubator, often the still air type, for the last three days of incubation and hatching.

The preceding information should be considered the "starting place", that when personal experience and a working knowledge of your incubator are added, will lead to successful hatchings.

Colony Breeding

Colony breeding is when a group of birds of the same species are housed in the one aviary for the purpose of reproduction.

Although there are a few records of successful breeding colonies of cockatoos, I feel in most cases the disadvantages of the colony system far outweigh the only apparent advantage of saving space. Of course, among the many disadvantages are the fighting and the aggression shown by the dominant pairs towards the rest of the colony.

Of the few successful colonies I am familiar with, all are housed in extremely large aviaries. In most instances of successful colonies only the dominant pair reproduces and the rest of the colony submit to their dominance. John Radford of New South Wales has the only totally successful breeding colony of cockatoos that I am familiar with, where three pairs of Major Mitchells are housed in a large aviary approximately 30 m (100 ft) long, in which all pairs rear young, most years. This situation is quite remarkable as many aviculturists in Australia and overseas will confirm just how aggressive male Majors can be when breeding. The maiming of young after fledging, by the other members of the colony, is always a possibility.

Other partially successful colonies that I am aware of are where a group of five Yellow-tailed Black Cockatoos are housed in the same aviary and one pair breeds successfully; also a pair of Galahs housed with several others rear young each season. Should a breeding colony of any species ever be established, be very careful not to upset the delicate balance that allows such a colony to function. For instance never add birds to the colony or take birds out, always remove young as soon as possible, never relocate the group into another aviary and never alter or relocate the furnishings, as there is only a hair line between harmony and disaster in most breeding colonies, which a minute change can turn into chaos.

Flock Breeding

My definition of flock breeding is where a group of disimiliar cockatoos or birds generally are held together for breeding purposes.

There are many records, mostly from zoos, where a pair of cockatoos which are part of large mixed collection housed in very large aviaries have bred successfully. In all instances these pairs have bred in spite of their environment, not because of it. Such pairs housed singly under reasonable conditions must become super breeding pairs.

As with colony breeding, the danger of the maiming of youngsters after they leave the nest is always present in flock situations.

Russell McAllister and Bill Connor of New South Wales recently informed me of their flock of 15 Black Cockatoos consisting of three pair of Red-tails, five Yellow-tails and two pairs of White-tails all housed in a large aviary of which, at the time of writing, two of the pairs of Red-tailed Black Cockatoos of the subspecies *naso* were incubating on one egg each, and a third pair had laid on the floor but refused to incubate. I will follow the progress of these breeding attempts with great interest.

Transportation

Recently I picked up a devoted breeding pair of Major Mitchells from a friend. They were placed in a weldmesh carrying cage 30 cm (1 ft) × 30 cm (1 ft) × 60 cm (2 ft) long and within two minutes the male had severely bitten the female on the foot, who then retaliated by biting the males beak and tongue, resulting in profuse bleeding from the mouth. I am ashamed to admit this indiscretion; there is no acceptable excuse, for I know better, but because the pair were so devoted and I was too lazy to look for a second cage, I took the chance, and only quick action in separating the pair avoided a disaster.

The moral of this story is never, under any circumstances, transport or hold cockatoos together in close confinement, even for a short period of time. Stress and fear generated by such situations can cause unprovoked and violent reactions by all cockatoo species and many other psittacines.

Over the years I have seen horrific injuries inflicted on unfortunate cockatoos who have been members of large groups bundled into wire crates or tea chests and shipped from one state to another for the pet trade. The welfare during transportation of birds of little value is only a minor consideration for many bird dealers and trappers, as, at their price, a percentage of losses is acceptable. Surely this situation must be unacceptable to the vast majority of people involved in the avicultural world, who undoubtedly must welcome some reasonable form of control. Of course valuable birds are given far different treatment during transportation.

Wildlife or Animal Welfare Authorities have, as yet, been unable to provide aviculturists or bird traders with a practical guide to transportation requirements for cockatoos.

The following points should be considered when transporting cockatoos:

1. Cockatoos must be boxed individually during transportation.
2. Only a minimum space should be provided for each bird as injuries often occur when frightened or stressed birds thrash about in too large a box or cage. My suggested formula for the size of a transport box is

the overall length of the cockatoo squared, to calculate the base area and the standing height of the bird plus 7.5 cm (3 in.) for the height, which provides adequate room for the bird to move without causing injury to itself. (See Figure 4.)

3. The box should be totally enclosed except for holes drilled in two sides to provide dimmed light and ventilation or half of the front should be weldmesh of adequate gauge. Enclosed boxes provide security for the bird, thus alleviating stress and injury.
4. Ventilator holes should be provided on at least two sides to allow the air circulation necessary if high temperatures are encountered.
5. Due to the chewing abilities of cockatoos, transport boxes must be constructed of a durable and strong material. Wooden boxes which must be made of a thick, hard timber to withstand a cockatoo beak, are heavy to carry and expensive to freight. Lightweight materials such as sheet aluminium can be made into excellent carrying boxes, but should be provided with a non-slip plywood false floor for the occupant's added comfort. Weldmesh wire cages of adequate gauge to withstand wire chewing, made to fit inside cardboard cartons which provide the necessary seclusion, are a light and cheap method of freighting cockatoos.

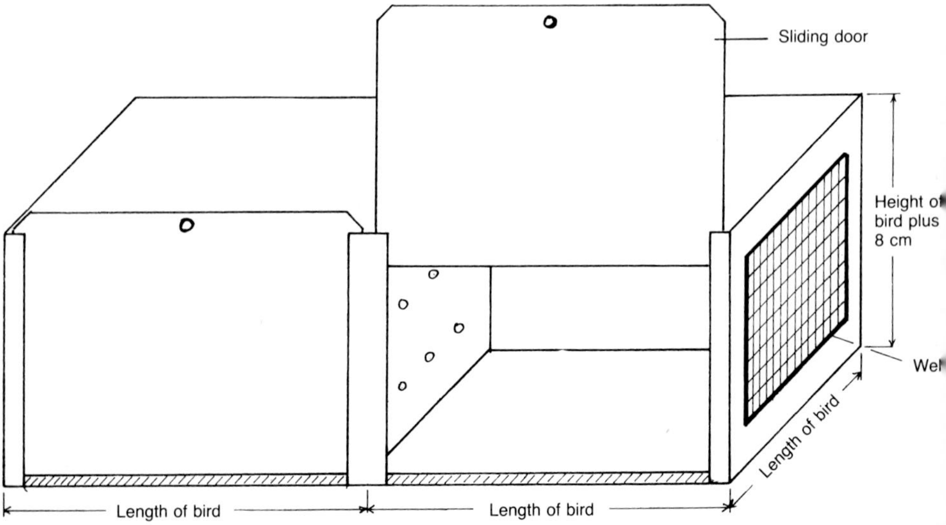

Figure 4. Dual-compartment transportation box.

6. When the duration of time spent in transportation is under four or five hours it is not necessary to provide water for the trip; however, water containers with the top edges turned over to form lips, to prevent the water waving out, should be fastened inside the box for longer journeys.

7. I find that seed spread on the floor of the box is more satisfactory than placing it in a container where it may not be found during the trip. A piece of apple will provide moisture when chewed by the birds.

8. Birds should never be transported during heatwave conditions unless in air-conditioned vehicles.

9. When birds are to be transported, unaccompanied, by air, rail or road, the box should be labelled "live birds", correctly addressed and a contact phone number stated, as well as provided with adequate food and water with a safety margin for the duration of the trip. The recipient must be notified well in advance of the pending arrival time.

Occupational Therapy

Cockatoos are intelligent, active birds who need to be amused and constantly occupied for their mental and physical well being. Boredom usually results in the development of vices such as feather plucking, screeching, wire chewing or the general demolition of their aviary.

Over the years various toys have been provided for cockatoos in an effort to amuse them and mellow their vices. Objects such as wooden cotton reels, empty tin cans and soft chewable perches have been common therapeutic toys for caged birds. Cockatoos housed in aviaries will sometimes enjoy similar items although most Australian aviculturists use branches from native trees and shrubs for chewing activities as well as various types of seed pods, cones, fruits, berries and green food which not only provide activity in the extraction of seeds, nuts, kernels etc., but nutrition as well.

Facilities that allow for the digging of holes is often enough to satisfy the members of the Corella family, Sulphur-crests and Galahs, but be warned, cockatoos that engage in this pastime must be wormed regularly.

However this activity therapy is applied or whatever the media used to apply it, ingenuity and perseverance is often needed to keep cockatoos happy and vice free in our aviaries. Of course in most instances each species or each individual for that matter, require a different approach to this occupational problem.

Acclimatisation of Black Cockatoos

Many aviculturists have experienced the anxiety and trauma generated when their Black Cockatoos refuse to eat after being transferred from one aviary to another. This is a common occurrence, obviously related to stress created by the new surroundings, which may last for a few days, usually without any ill effects.

I would suggest that whenever Black Cockatoos, or any other large psittacines for that matter, be relocated, several temporary food containers be placed in strategic positions such as fixed on to or adjacent to, the perches so that food can be seen and partaken with a minimum of effort on the part of the stressed bird.

If the bird in question is a new acquisition, always enquire of the previous owner exactly what it normally eats and provide the same, at least until it settles in.

Immature Black Cockatoos up to the age of 18 months or so, will often refuse to eat when stressed due to a move from one place to another (or due to illness), and often revert to begging for food just as they had to their parents, months or even a year earlier. Supplementary hand feeding of softfood from a spoon or syringe may be necessary for a few days, until confidence is restored and self feeding resumes.

DISEASES

James Gill, BVSc, MVM, MACVSc

Successful Cockatoo keeping and their subsequent reproduction is a function of the proper selection of good birds in the first place, followed by correct housing, good nutrition and management programmes including effective disease control.

Cockatoos as a group are subject to the same wide range of diseases as are other birds.

Any bird can get one of two forms of disease — infectious or non-infectious disease. For birds to become ill with an infectious disease they must be exposed to a sufficient number of the infective agents (for example, virus particles or worm eggs) to overcome the bird's natural defence mechanisms, that is its immune system. Non-infectious diseases include such things as fractures of the legs and wings, egg binding, starvation, vitamin and mineral deficiencies. Some of the non-infectious diseases can predispose the birds to infectious diseases. The strength of the bird's immune system can be lowered by such things as cold, damp conditions, deficiencies in vitamins and minerals, lack of sufficient quantities of food to eat and constant stress from other birds, dogs, cats, mice, rats or people including their owners.

Therefore, to minimise the chances of birds getting sick we must:

1. Reduce the birds' access to bacteria, viruses, parasites, etc. This can best be done by a month's quarantine on all birds coming into the collection and by maintaining good hygiene for all birds in the collection.

2. Make sure the birds' immune systems are working well. This is best done by ensuring that they are housed and fed well.

We should all strive to make preventative medicine the more important aim in our hobby and curative medicine a secondary role.

A large percentage of disease problems in any livestock system can be eliminated by good management. The system of housing, feeding and management as outlined by the authors in this book will go a long way towards eliminating disease problems.

Proper health management is at least partly a function of the proper planning, design, construction and maintenance of the aviaries themselves. Several important features should be highlighted here as they relate to disease control. It is essential that fresh water be provided daily. For the aviculturist who has more than a few aviaries, to do this consistently an automatic watering system is an advantage. (See Management chapter.)

Concrete floors are considered by me to be essential. They enable aviaries to be cleaned thoroughly. To assist in this routine cleaning and drying, they should slope gently towards a drainage system.

Walls should be of single thickness construction so as not to provide hiding places for mice or other vermin.

There should always be two doors between the birds and the outside of the aviary so as to prevent unnecessary losses due to escapes.

The divisions between aviaries should be of solid material or double wire to prevent traumatic injuries from neighbouring birds. Perches should not be placed over food or water containers.

Food and water containers should be easily cleanable.

All new wire should be weathered or treated with vinegar and a wire brush to remove excess galvanising and oxide. Heavy metal poisoning from new wire is a common yet easily avoidable disease.

Remember all that any aviary has to do is provide a secure hygienic environment that gives the cockatoo the ability to exercise. Many other adornments are usually only for human aesthetic needs.

Quarantine

The isolation or quarantine of all incoming cockatoos for a period of at least 30 days is the first step in a successful disease control programme. This period enables the proper examination of the birds and observation of their behavioural patterns, particularly their eating habits.

During this period the cockatoos should be examined for any deformities, if they were not examined prior to purchase. They should be treated for any external and internal parasites.

The most important feature of quarantine is to prevent the introduction of new infectious agents into the collection. No cockatoo should be removed from quarantine area to the general collection unless the aviculturist is completely happy with its health.

With valuable birds or valuable collections it is prudent to weigh the birds regularly during quarantine and to do direct faecal smears and faecal flotations at least twice looking for internal parasites.

When the legal importation of avian genetic material starts into Australia it would also be wise for aviculturists to quarantine birds once they leave the government quarantine facility. These birds may be introducing new infectious diseases that are not tested for by the government authorities. I would recommend the use of sentinel birds in your private quarantine areas as a simple effective way of checking for any potential problems. The cockatiel would be ideal for this purpose alongside any cockatoos. Any bird, be it a new arrival or sentinel bird, that dies in your quarantine should have a thorough post mortem performed on it.

Another major advantage of quarantine, which is often ignored, is that it provides an observation period to note the birds' behavioural patterns particularly their eating habits. Many newly acquired birds, especially freshly trapped wild birds, die of starvation because they are not familar with the food offered and refuse to eat it. Black Cockatoos can be a particular cause for concern when they are moved and may take several days to start eating again even when they are familar with the food. By keeping newly acquired birds by themselves or in pairs you can observe if they are eating the diet provided and adjust the diet accordingly. This can avoid many unnecessary deaths.

Cockatoos, because of their size, require large cages for quarantine. 1.2 m (4 ft) cubed suspended wire aviaries have worked well for me holding pairs of cockatoos for several months at a time. The quarantine area should be isolated from the other aviaries and the birds should be fed and watered after the other birds in the collection have been cared for.

Diseases of the Skin and Feathers

Traumatic injuries to the skin, beak and toes are quite common but should be largely prevented by good management. Cockatoos should be transported as individuals and not in pairs or groups. There should always be solid partitions or double wiring between aviaries and carrying cages. Particular care should be noted of pairs during the breeding season as males can be aggressive to their mates. Most wounds resolve without complications but toe injuries in particular require vigorous antibiotic therapy to prevent the spread of infection. Beak injuries can be quite complicated and have a devastating effect on the bird especially if they lose the top beak. Some of these injuries require surgery which in my hands at least has a guarded prognosis. It is surprising how many parrots learn to eat without a top beak and providing they can survive the first few days do very well. Bleeding can be a problem with some of these injuries and is best controlled by restraining the bird and applying direct physical pressure to the area for two or three minutes. If this is not successful Condys Crystals (potassium permanganate) can be applied to small wounds to achieve haemostasis. Severe wounds will require veterinary attention.

Bumblefoot is an infection of the sole of the foot of cockatoos; it is particularly common in pet birds housed in unhygienic cages with ill sized perches and fed faulty diets. I have also seen it commonly in aviary Major Mitchells. Many of the milder cases respond well to antibiotic and injectable vitamin A therapy. The vitamin A appears to improve the health of the tissues in the foot. The severe cases have a poor prognosis for complete recovery.

Mutilation of feathers can occur in a number of ways. The most common form is self mutilation, where birds, through boredom, sexual frustration or an irritation of the skin or deeper parts of the body start chewing their own feathers. You should, with the help of your avian veterinarian, try and find the cause and remove it. With the psychological problems, most notably with Gang Gang Cockatoos this is easier said than done. With breeding birds we routinely pluck the damaged feathers and adjust the environment to try and overcome the boredom. In pet birds if modification of the environment is unsuccessful then the use of injections of medroxyprogesterone acetate by experienced avian veterinarians has proved to be successful in a large percentage of cases. Some adult pairs will also pluck one another.

Parents will also pluck their chicks in the nest and in severe cases will damage wings and toes in the process. The simplest solution to this problem is to remove the young for hand rearing.

Psittacine Beak and Feather Disease is one of the most common and frustrating diseases facing keepers of cockatoos. In our practice in Sydney we see large numbers of Sulphur Crested Cockatoos affected with this disease in their first year of life. Most of these birds have been illegally taken from the wild as nestlings. Smaller numbers of the other cockatoos are also seen with the disease including aviary bred birds. Drs Pass and Wylie, working at Murdoch University in Western Australia, have succeeded in transmitting the disease by inoculation of susceptible young birds with the virus particles. The virus itself has not been cultured. Internationally there appears to be differences in the pathology and aetiology of the disease. It may well be that different aetiologies produce the same clinical picture. In time this confusion will no doubt be corrected as our understanding of the disease process in avian skin and feathers improves.

The features I look for are pinching or narrowing of the feather shaft as it leaves the feather follicle. This may be accompanied by haemorrhage in the feather shaft. These feathers are therefore brittle and fall out easily and the bird can end up completely naked. In Sulphur Crested Cockatoos the feathers of the crest are often affected early in the disease process. The birds are often very dirty in appearance. Skin biopsies may be necessary to confirm the diagnosis in mild cases. Deformities of the beak (beak rot) are a common feature of the disease in Sulphur Crested Cockatoos and Gang Gangs. At this stage the disease is incurable and the cockatoos in contrast to some other species of parrots do not recover. It is an infectious disease and affected birds should be removed from a breeding collection.

Skin tumours or more correctly lipomas (fatty tumors) under the skin are very common in Galahs and Sulphur Crested Cockatoos, especially those kept as pets. They can be prevented and reduced in size by vigorous dieting. Once they become ulcerating or start affecting the movement of the bird they should be removed surgically.

Diseases Associated with Reproduction

Egg binding is a common problem with a complex aetiology. Calcium deficient diets, obesity, chilling, lack of exercise and infections of the oviduct can all lead to egg binding. I find the majority of cases respond well to an increase in environmental temperature and administration orally of calcium (Calcium Sandoz Syrup) 2 mls every two hours for cockatoos until the egg is passed. In very depressed hens full suppportive therapy is instituted, including intravenous fluid therapy. In birds that fail to respond to this regime and who are deteriorating, surgical therapy is considered. Radiographs may be needed to confirm the diagnosis.

Fortunately infertility is not a major problem with any of the Australian Cockatoos with even very imprinted hand reared male Black Cockatoos being known to fertilise eggs. Most problems associated with failure to breed can be traced to management problems related to housing, nutrition and incompatiblity of the breeding pairs. Full clinical workups including semen evaluation and endoscopic examination of the gonads in the breeding season are essential in evaluating non-breeding birds. The use of mammalian hormones in birds in a blind attempt to stimulate breeding is of doubtful value and may be dangerous in the long term.

Pathological fractures are common in young cockatoos, both parent reared and hand reared. They are the direct result of feeding calcium deficient diets. They are seen commonly at two stages in the growth of young cockatoos. Firstly during the rapid growth phase at two to four weeks of age and also at fledging when they normally have a reduced food intake to lose weight to assist with flying. Any marginal deficiencies are likely to show up at these times as fractures of the tibiotarsus firstly and then followed by the wings (humerus and radius and ulna) as they attempt to drag themselves around the nest or box.

Once the fractures occur, and they are usually multiple, I treat them with external fixation and adjust the diet. Calcium Sandoz Syrup is readily available and a quickly utilised source of calcium. Young cockatoos are given 1 ml three times a day. Recently I have improved results using very large doses of injectable Calcium Sandoz. More importantly, the breeder should adjust the diet for the next breeding season. Many cockatoos resist eating cuttlefish or mineral blocks and it is often easier to add Calcium Sandoz Syrup or calcium carbonate powder to their fruit or sprouted seed. If birds could be induced to eat acceptable pelleted rations it would overcome many of these problems. (See Diet chapter).

Many hand rearing diets are also deficient in calcium and this becomes evident particularly when very young chicks are being reared. I have personally hand reared many large parrots and cockatoos using the formulas as outlined in the Diet chapter and have not had pathological fractures develop.

I personally consider nutrition to be the single most important limiting factor in the reproductive performance of cockatoos in captivity. In recent years tremendous advances have been made in our understanding of housing, management and disease control in captive birds but unfortunately our knowledge of the dietary requirements of the many species of cockatoos is lagging behind.

Generalised Infectious Diseases

A wide range of bacteria have been reported to be pathogenic in cockatoos. Most of these are best controlled by good hygiene and quarantine. When a bird becomes infected it generally involves more than one body system at a time. If the illness is detected early then with improving treatment regimes including injectable antibiotic and fluid therapy many birds can be saved. With valuable individual birds or when outbreaks occur in a collection then bacterial culture and sensitivity testing is essential to maximise the chances of successful treatment.

It should be remembered that some of these bacterial diseases, most notably Salmonella and Chlamydia, are potentially dangerous to the human owners of the birds.

All bacterial diseases require experienced veterinary treatment to maximise the chances of a successful outcome.

Nephritis or infections of the kidney are a common primary disease of cockatoos especially those kept as pets in unhygienic conditions. Bacteria from dirty perches invade abrasions or small cuts in the birds feet and can be picked up by the blood supply and deposited in the kidneys. The birds become ill very quickly and are often vomiting and have profuse urine output. Depending on the degree of damage done to the kidneys they respond well to antibiotic and fluid therapy. I routinely recommend to owners of pet birds that they have two sets of perches and rotate them week about and clean the set not in use during the week and leave them out in the sun to dry.

Fungal Diseases

Aspergillosis is generally a chronic disease in cockatoos involving the respiratory tract, especially the lungs and air sacs. It is not a common problem in southern Australia.

Candidia can become a problem of the upper digestive tract of young hand reared cockatoos. We routinely use Mycostatin in hand reared parrots that have any slowing of food passage time or in babies that start vomiting. The fungal elements can be seen on stained samples of the crop contents.

I also recognise an entity in adult Black Cockatoos with vomiting that we believe is caused by Candidia. Numerous candidia are seen on stained samples of this vomit or on material from crop washes. The differential diagnosis in these cases includes trichomoniasis, nephritis and heavy metal poisoning. They are treated with Mycostatin orally, antibiotics systemically and vigorous systemic fluid therapy.

Parasitic Diseases

Giardia, which apparently causes significant losses in California amongst Cockatiels, is not a problem in Australia at this point in time.

Ascarids or roundworms are a major problem in all parrot species. They can be controlled by the use of concrete or wire floors and by worming all birds as they come into quarantine. In Australia Panacur 2.5 (Fenbendazole 25 g/l) is the most commonly used worming drug for cockatoos. It is not registered for use in birds and the user must assume responsibility for its use. The following dose has been used with safety: 0.4 ml/100 gm bodyweight with a maximium of 2 mls given to the larger cockatoos. It should not be used when birds are moulting as it has been incriminated in causing feather abnormalities.

Capillaria have not been a major problem in cockatoos in our practice. However, they are more difficult to erradicate. I use Panacur 2.5 at 0.2 ml/100 gm bodyweight daily for three days.

Tapeworms are occassionally seen in wild caught cockatoos and can be treated with Droncit tablets (Praziquantel) at 5 mg/kg.

Surgical Sexing

Surgical sexing is essential to get pairs of Corellas and to accurately determine the sex of many young Black Cockatoos. It is relatively safe, accurate and a fast way of determining the sex of these birds.

GENUS PROBOSCIGER

Plate 2: Palm Cockatoo *Probosciger aterrimus.*

PALM COCKATOO

(Probosciger aterrimus)

Derivation
Probosciger
  proboscis — Greek for nose
  gero — Greek for carry
aterrimus — Latin for very black

Description
  Length 60 cm
  Weight 1000 g (males are usually considerably heavier than females)
  Appearance as in Plate No. 2

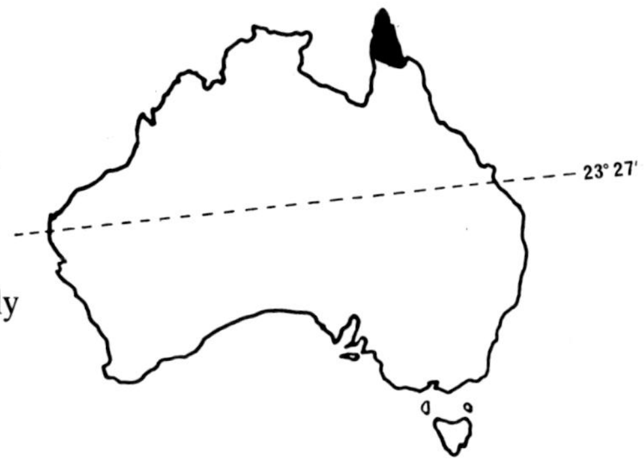

Classification

The Palm Cockatoo is the only member of the genus Probosciger (monotypic) in which Forshaw records three subspecies. It is distinct from all other Cockatoos with its massive protruding upper mandible and smaller lower mandible, which do not fit compatibly together when the beak is closed (unlike all other Cockatoos). This allows the small red, black tipped tongue to be visible. My first experience of this unusual beak structure occurred many years ago, when I agreed to help Bob Lynn to worm his Palm Cockatoos. I could not believe the total independence of the upper and lower beaks. The bases were so far apart and their shape so incompatible, that the impression given was that they did not belong to the same bird. The large red fleshy area of the mouth and throat which separates the mandibles is broken only by the base of the small dexterious tongue. The mandible and mouth structure is unlike that of any other pssittacine. This cockatoo's cheek patches are similar to the bare cheek patches of the Macaw family and gave rise to the name of Macaw Cockatoo, which was often used for this species in the last century. This possibly reflects parallel evolution between the two. Like the Macaw the cheek patch is capable of changing its depth of colour, becoming a deeper red in the mating season and when the bird becomes excited or agitated, in other words it blushes. The crest is more extensive than any other cockatoo and consists of long narrow feathers which curve backwards over the head. Unlike other Black Cockatoo species the cere is feathered and unlike all other cockatoos the thighs are bare of feathers.

Earliest Report

This cockatoo was described in 1707 by the Dutch author S. Van der Meulen and first figured, rather poorly, by Pelvo Scheuk as *Corvus indicus* in a small book published in Amsterdam. The species was more adequately described and portrayed by Edwards in his Gleanings of Natural History Vol. 7 page 229 in 1764. Gmelin scientifically categorised this species in 1788. Macgillivary discovered the species on Cape York when the HMS *Rattlesnake* visited there in 1847 and his notes were subsequently published by Gould in his "Birds Of Australia" Supplement 1851 and in his Handbook to the Birds of Australia Vol. 2 page 27, 1865.

Range Habitat and Field Notes

The Palm Cockatoo ranges through West Irian, Papua New Guinea and adjacent islands and the Cape York Peninsula in Australia. There are three recognised subspecies:

Probosciger aterrimus aterrimus which is the nominate subspecies and the one found in Australia.

Probosciger aterrimus goliath which is larger than *P. a. aterrimus*. Generally speaking this subspecies ranges from the western Papuan Islands and from Vogelkop in West Irian eastwards through the central areas to eastern Papua.

Probosciger aterrimus stenolophus is similar to *P. a. goliath* but has narrower crest feathers. It ranges from Japen Island in Geelvink Bay through West Irian eastwards across northern New Guinea to Collingwood Bay in Papua.

P. a. aterrimus ranges from Aru Islands and approximately through southern West Irian, eastward to the Gulf of Papua and to the Cape York Peninsula in northern Australia.

The exact range of this cockatoo is yet to be determined even on the Cape York Peninsula, where Forshaw credits this species with a far smaller range than do other authors.The RAOU "Atlas of Australian Birds" provides the most recent information on the ranges of Australian birds and it suggests that the Palm Cockatoo inhabits almost the whole of the Cape York Peninsula as far south as the Princess Charlotte Bay region in the east and Cape Keer-weer in the west.

On Cape York Peninsula this cockatoo inhabits the rainforest edges and adjacent eucalypt forest. It often moves considerable distances from rainforest areas through the drier eucalypt forests bordering water courses in search for food.

My first sighting of the Palm Cockatoo was in low stunted eucalypt forest adjacent to a fast flowing creek. We had stopped at the creek for a lunch break, when the unmistakable whistling call was heard. Three birds were found perched in low trees no more than 5 m above the ground. Upon our approach one bird moved off, flying slowly through the low trees, rather than above them.The other two remained and allowed us to observe them from 25 m away. They were obviously aware of our presence but were prepared to tolerate us provided we ventured no closer. The birds were perched in separate trees and casually preened their feathers or chewed at branches and foliage, they were not feeding but appeared to be resting in the canopy of the low forest to avoid the heat of the day. After allowing us to watch them for some 15 or 20 minutes, they flew off through the trees, uttering a couple of subdued calls.

Although we sighted Palm Cockatoos on several occasions while on this trip to the Iron Range, usually in eucalypt forest close to water courses, or on the edge of the rainforest, they never again allowed us such a close approach as on that first sighting. Another sighting of this species which has left a picture in my mind that I am sure I will carry to my grave, was of a pair that flew noisily along a creek just above the height of the bordering trees. It was late afternoon and the birds were highlighted by the sun which was low in the sky. They passed almost directly overhead giving my wife and I a wonderful view of them in flight. The male's red cheek patch was absolutely vivid, far more intense in colour than any other Palm Cockatoo I have ever seen. When in flight the Palm Cockatoo carries its head low giving a straight backed appearance while it flies with full, strong, slow wing beats.

In past literature there has been occasional reference to the Palm Cockatoo's ability as a drummer, but I had no idea of the full significance of these references. That was until I viewed the video which was shown in conjunction with Peter Chapman's informative lecture on Black Cockatoos, at the 1987 Australian Federation of Aviculture's Convention held at Gosford, New South Wales. This has now been released for sale to the public as one of a series called "The Land of Parrots" and makes excellent viewing. The segment dealing with the Palm Cockatoo was superb. Besides showing a number of Palms drinking on the bank of a river, there was a section of the film which showed a magnificent male bird perched on a dead hollow tree, holding a stick in his claw. The stick appeared to be about 30 cm (12 in.) long and 35 mm (1½ in.) in diameter. With this stick he excitedly drummed on the hollow trunk creating some considerable sound. As the film progressed he changed the stick to the other foot and then continued to drum. It is comforting to know that this brilliant performance by such a magnificent specimen was captured on film for all to enjoy.

Diet in the Wild

Palm Cockatoos feed on fruits, nuts, berries, seeds and vegetable matter. Some early reports suggest that like the other Black Cockatoos, they also feed on larvae of wood boring insects. Forshaw discounts this because of insufficient evidence. This species has often been observed feeding on the ground from fallen fruits, nuts, seeds, etc. Seed from the fruit of the Pandanus Palm has long been considered a favourite food.

Breeding in the Wild

The nesting sight is usually a vertical or near vertical hollow limb or trunk of a tree, often in dead timber. Nesting cavities may vary considerably from 30 cm (1 ft) to 60 cm (2 ft) in diameter and range in height from 3 m (10 ft) to 16 m (50 ft) above the ground. The depth of the nest from the entrance hole may vary from 60 cm (2 ft) to 2.5 m (8 ft).

It is prepared by both the male and the female on a bed of chewed up sticks, which are carried to the nest in 30 to 45 cm (1 to 1.5 ft) lengths. They are dropped down the hole and then chewed to splinters by both birds. The depth of this bed of chewed up sticks may vary from several centimetres to two or more metres in depth. It is considered to be a means of combating the effects of torrential rains from flooding the nest by providing an absorbent base.

The nesting period is usually from August to January with the most favoured months appearing to be August, September and October. One oval white egg is laid and incubated by both female and male. Incubation and fledgling periods do not appear to have been recorded in the wild. Young birds are fed by and remain in the company of their parents for many months after fledging. Immature birds show yellow spotting on the lower frontal body feathers which produces a faint barring effect.

Aviculture

This species was first exhibited in Europe by the Director of the Zoological Gardens of Amsterdam in 1860 and in England by the Zoological Society of London in 1861. Although available in Europe and England around this period, it was considered to be a pet bird, hence early avicultural information is non existent. During the 1880's this bird commanded the prohibitive price of £25 to £30 each, which indicates how few must have been available at the time.

The earliest report of a Palm Cockatoo in captivity in Australia appears to be of the one survivor of two babies taken from nests on Cape York Peninsula in 1912 by the renowned ornithological collector W. R. McLennan. Both babies were sent to Dr E. A. D'Ombrain of Sydney. One died the morning after arrival and the other was reared. At the time Dr D'Ombrain's article relating to this event was published in the "Emu" Volume 33 part 2, pages 114-121, October 1933, the bird was 21 years old.

In the USA the following near breeding successes have been recorded. San Diego Zoo recorded in the Avicultural Magazine 1951, page 69 that in May 1948, two fertile eggs were laid but were deserted by the parents. In August 1950 one of the two eggs laid was hatched but the chick was rejected by the parents. The interesting point about this report is that it seems to imply that this pair of Palm Cockatoos laid two eggs per clutch.

Frank Buck wrote in the Avicultural Magazine 1944 page 163 of a J. D. Putman who had a young Palm Cockatoo in the nest. David West, *in litt.*, 12 December, 1949, states that in fact Putman hatched but did not rear the young Palm. David West also stated that in 1944 Sheffler of Arizona reared a Palm Cockatoo.

The lack of documentation places serious doubts on this being a parent reared breeding. This opinion must also be shared by the U.S. Justice Department, who in 1985 flew one of their officers to Australia, to interview Mr Bob Lynn on his world first and subsequent breedings of the Palm Cockatoo. The officer was gathering authentic information to be used in subsequent legal proceedings relating to the importation of Palm Cockatoos into the USA.

What has been acclaimed in Australia as the world first parent reared breeding of the Palm Cockatoo was achieved by Bob Lynn of Sydney, New South Wales, Australia in 1968. After extensive investigation I am convinced that it is a world first breeding. The following is Bob's report of the events leading up to and of his world first breeding.

"I purchased two Palm Cockatoos, which were supposed to be a pair, from Dave Bush who acted as an agent on behalf of Sir Edward Hallstrom. An aviary had previously been erected to house these birds, which was 12 m (40 ft) long, 3.6 m (12 ft) wide and 3 m (10 ft) high. The aviary faced east and had a 2 m shelter on the western end. The southern side was totally enclosed and the eastern end was partially enclosed with fibreglass.

Apart from the roofed shelter on the western end, the aviary was roofed for one third of the eastern end with fibreglass.

"The two Palms were similar to each other with one being slightly larger. They were obviously not compatible, avoiding each other at all times and were always found at opposite ends of the perch, if not the aviary. This made me doubt that they were in fact a pair. When their attitude towards each other had not improved after a few weeks I phoned Sir Edward to discuss the problem. He also felt that they weren't compatible and said that he had only one other hen available, which was very small, but if I liked to collect her I could try her with the other two birds to see what transpired.

"A few days later I collected the hen and placed it in the aviary with the other two birds. The largest of the two original birds immediately flew down to the hen, who was still on the aviary floor, and displayed to her. She appeared too weak to fly and remained on the aviary floor. After this display to the new bird, this now obvious cock bird flew straight at his original companion and attacked him so viciously that I am sure had I not removed him he would have been killed. The rejected bird was later returned to Sir Edward Hallstrom. Besides being a small bird the hen also had a deformed beak. This, added to her reluctance to fly, prompted me to phone Sir Edward again, to talk about the deformity and to enquire if I could exchange it for a better bird. I was told that there was no other Palm Cockatoo available and it was this pair or nothing, so I had no choice but to do the best I could with what I had, like it or not. He did offer to trim the beak for me but I declined the offer.

"I knew Sir Edward Hallstrom's Palm Cockatoos had been collected in New Guinea, many years before by Dave Bush, so I contacted Dave and asked if he had brought back a bird from New Guinea with a deformed beak. Dave could not remember the individual birds after 30 years had elapsed, but assured me that no bird would have been brought to Australia unless it was in perfect physical condition as there had been plenty of birds to choose from. This information suggested to me that perhaps with a good diet and plenty of chewing material, such as Eucalypt and Hakea branches, the deformity could be improved.

"A few months later, while in my aviaries feeding, I noticed something different about the hen Palm, who by this time was flying strongly. On closer examination I found that the overgowth on the top mandible had come off, leaving the beak perfectly normal. The constant chewing and the improved diet had achieved what I am sure beak trimming could not.

"The nesting log provided stood directly on the ground in a vertical position and was 1.5 m (5 ft) high and 37 cm (15 in.) in diameter with the entrance and access straight down the top of the log. Rotting wood and dirt taken from old stumps in the bush was placed in the log to a depth of 45 cm (18 in.). Not long after the hens beak had returned to normal the pair started to show interest in the log and began to fill the hollow with chewed up pieces of Hakea and Eucalypt branches. The Hakea branches were used because I had discovered that Palms had a liking for Hakea nuts. The hollow was filled by both the male and female to within 45 cm (18 in.) of the top.

"Soon it became apparent that the hen was coming into egg and then to my horror it was obvious the hen was egg bound. I massaged the vent with paraffin oil and also gave her a dose of paraffin oil to the beak, but she showed no improvement and could barely drag herself around the floor of the aviary. Then to my relief, after three days, on 12 February, 1968, she laid the egg in the nesting log but was still that sick that she left the egg and refused to incubate. The cock then proved himself to be an exceptional bird for he commenced to incubate on his own and continued to do so for one week, day and night, until the female recovered. She then started to share incubation duties with the male during the day and carried out all incubation herself throughout the night.

"The cock Palm proved to me he was one bird in a million and I feel fortunate that this first egg was laid in the hottest months of the year. The hen would have been lucky to survive this ordeal in the winter months. The hen never had difficulty in laying her eggs after this first incident.

"After an incubation period of 33 days the egg hatched on 16 March, 1968. The chick hatched naked with closed ear orifices and with the upper mandible noticeably larger than other Black Cockatoo babies. Pin feather development started to appear between three and four weeks of age. Both parents fed the chick in the log. The young Palm Cockatoo left the nest on 5 June, 1968, 81 days after hatching. It was a fine strong youngster 7.5 cm (3 in.) longer than its father, with a much smaller horn-coloured top mandible grading to black at the base. The cheek patch was a pale pinkish red and there was yellow spotting on the belly area.

"The youngster was fed by both parents for three to four months after fledging and neither parent showed any sign of aggression towards their offspring prior to it being removed from the aviary at six months of age.

"My diet fed to the Palm Cockatoos while they were rearing the young was a basic sunflower seed mix together with peanuts, almonds, brazil nuts (which were shelled and cut into pieces to try and eliminate wastage), Hakea nuts and silverbeet of which they were very fond."

Sexing

Palm Cockatoos can be difficult to sex because of the variation in size between individuals, some males being smaller than large females. Normally hens are smaller in the head and the upper mandible is noticeably smaller than in the male. Surgical sexing can save a lot of wasted time if there is any doubt about a bird's sex in this species.

Display

The courtship display observed by Bob Lynn in his aviaries was simple and similar to other cockatoos. The male stretches high, raises his crest, his cheek patches deepen in colour and he calls excitedly as he advances towards the hen.

Nests

The nesting site found to be most favoured in Bob Lynn's aviaries, both by the original pair and by the aviary bred young, was a vertical log about 1.5 m (5 ft) high and 37-45 cm (15-18 in.) internal diameter with an entrance gained down the top of the log. Natural looking irregular shaped entrances are the most preferred and a substantial perch located alongside the entrance of the log was found to be favoured.

At Taronga Zoo, Sydney, in 1971 the then Curator of Birds, Kerry Muller, had a pair of Palms breed in a similar sized log as previously described. It was filled to a depth of 30 cm (12 in.) with natural wood chips and the pair added a further 5 cm (2 in.) of chewed up sticks. Unfortunately the baby was lost as a result of psittacosis, some time after being taken for hand rearing.

In 1973 a second pair at Taronga Zoo nested successfully in a larger vertical log which was stood on the ground and measured 2.3 m (7 ft 6 in.) high and 75 cm (2 ft 6 in.) in diameter. This chick eventually fledged after one hundred days, but it was a weak youngster and although it did improve considerably after flying, unfortunately died some months later.

Nesting and Hatching

Palm Cockatoos lay one egg and usually commence incubation within 24 hours of it being laid. Both male and female incubate with the female carrying out all the night duties. Bob Lynn recorded two 33-day incubation periods and one 34-day incubation period. One 33-day incubation period occurred during February-March and the other two periods were October-

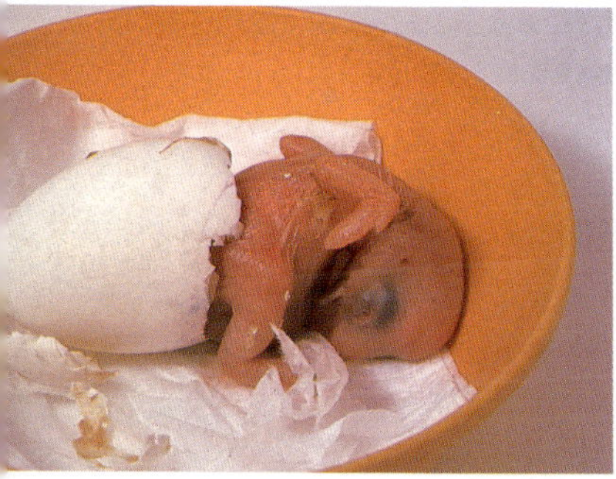

wly hatched Palm Cockatoo at Pearl Coast
o, Western Australia.

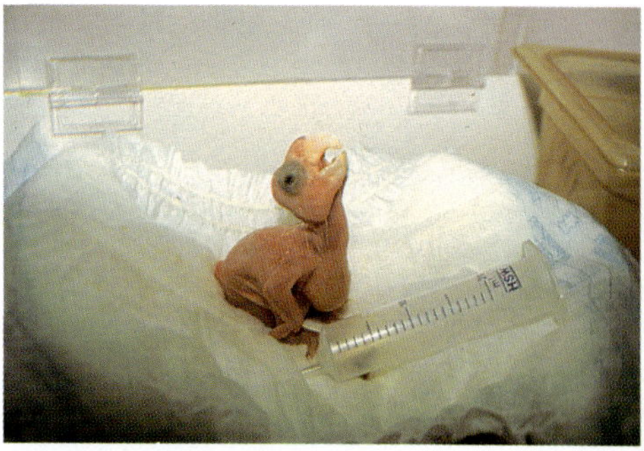

B Palm Cockatoo nestling 10 days old at Avicultural Breeding and Research Center, Florida, USA.

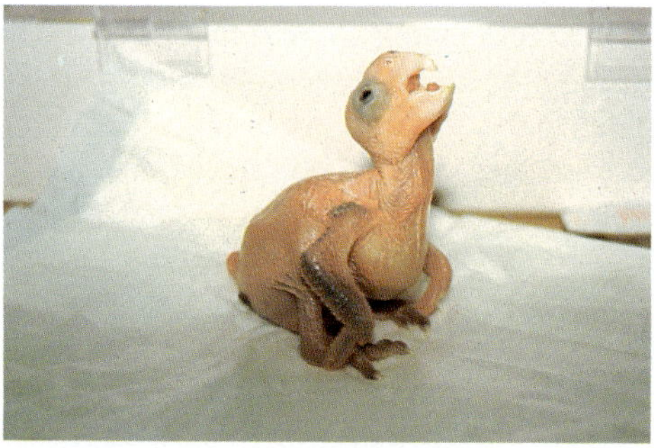

C Nestling 12 days old photographed at Avicultural Breeding and Research Center, Florida, USA.

stling 22 days old photographed at
icultural Breeding and Research Center,
rida, USA.

E Nestling 118 days old photographed at Avicultural Breeding and Research Center, Florida, USA.

Plate 3: Palm Cockatoo Nestlings.

November, so obviously the time of the year had no relationship to the one-day variation in the recorded incubation periods.

The chicks hatch naked with sealed ears and a noticeably large upper mandible in Bob's aviaries. Pin feathers appear from three to four weeks of age. Bob recorded fledging periods of 81 days and two 78-day periods.

Kerry Muller recorded that the Palm Cockatoo chicks hatched at Taronga Zoo were hatched naked and developed a sparse grey down about two weeks of age.

When young Palm Cockatoos fledge, they are similar to adults with a smaller horn-coloured upper mandible which is blackish at the base. The lower mandible is pale grey. The feet and the legs are pale grey and the eye ring is a paler shade as is the red cheek patch.

Bob Lynn had chicks hatch in March and twice in November and Taronga Zoo had them hatch in November and April. Bob recorded eggs laid in January, February, March, April, July, October, November and December over a four-year period. When we consider these records I feel that it is reasonable to assume that Palm Cockatoos may be expected to nest at any time of the year in Sydney and areas with a similar climate. Besides his world first breeding documented previously, Bob Lynn successfully reared Palm Cockatoos on two other occasions. One fledged on 16 February, 1970, and the other on 28 January, 1972. All were exceptional youngsters and all are still alive and thriving today. The first bird reared proved to be a male and the second and third birds were both females. The first female reared was paired to its brother and laid its first egg at four and a half years of age. Unfortunately Bob was not successful with this pair during the several years they remained in his possession, as the male insisted on taking the eggs from the nest and dropping them from the perch. Had more birds been available perhaps compatible pairs could have been established and better results obtained.

Various authors have published cockatoo incubation periods that are ascertained from artificially incubated eggs, which I feel, although useful, must be treated as guides only. I have found that slight temperature or humidity variations during artificial incubation can influence the length of incubation periods considerably. Likewise fledging periods derived from hand reared birds are totally unreliable, being too dependant on the nutritional content of the diet fed to the youngsters.

Mutations

An Albino Palm Cockatoo was recently acquired in Indonesia and sold by a Singapore bird dealer to an unnamed aviculturist.

GENUS CALYPTORHYNCHUS

GENUS CALYPTORHYNCHUS

This is a diverse genus in which all members are sexually dimorphic and unlike all other Australian Cockatoo genera, only the female incubates.

The genus is divided into two groups, the Yellow and White-tailed species in which sexual dimorphism is slight, and the two Red-tailed species in which the sexual dimorphism is pronounced.

Most recent authors follow the fashionable trend of lumping species together and in this genus the Yellow-tailed Black Cockatoo *(Calyptorhynchus funereus)* and the White-tailed Black Cockatoo *(Calyptorhynchus baudinii)* are usually classed as members of the same species.

Current information suggests that the longer billed subspecies of the White-tailed Black Cockatoo *(Calyptorhynchus baudinii baudinii)* should be given separate species status. Until this is clarified I shall divide this genus as follows, in what I consider to be a simple and perhaps avicultural approach to what has become a complex genus.

Yellow-tailed Black Cockatoo *(Calyptorhynchus funereus)*
Two subspecies: *Calyptorhynchus funereus funereus*
Calyptorhynchus funereus xanthanotus

White-tailed Black Cockatoo *(Calyptorhynchus baudinii)*
Two subspecies *Calyptorhynchus baudinii baudinii*
Calyptorhynchus baudinii latirostris

Red-tailed Black Cockatoo *(Calyptorhynchus magnificus)*
Four subspecies *Calyptorhynchus magnificus magnificus*
Calyptorhynchus magnificus macrorhynchus
Calyptorhynchus magnificus samueli
Calyptorhynchus magnificus naso

Glossy Black Cockatoo *(Calyptorhynchus lathami)*

YELLOW-TAILED BLACK COCKATOO

Plate 4: Yellow-tailed Black Cockatoo *Calyptorhynchus funereus* — Male.

YELLOW-TAILED BLACK COCKATOO

(Calyptorhynchus funereus)

Derivation:

Calyptorhynchus:
calyptos — Greek for hidden
rhynchus — Greek for beak
funereus — Latin for funeral or black

Description:
Length 65 cm
Weight 800 g average (females are usually heavier than males)
Appearance as in Plates Nos. 4 and 5

Classification

I find there is a distinct line dividing the Yellow-tailed and White-tailed Black Cockatoos from the Red-tailed species, although all are members of the one genus. Some of the prominent differences are the variation in the extent of sexual dimorphism, the coloured periophthalmic eye ring of the male Yellow and White-tailed species and that two egg clutches are often laid by the Yellow and White-tailed species whereas one egg clutches are the norm for the Red-tailed species. The Yellow and White-tailed species have a more protruding and pointed upper and lower mandible, which appears to be more suited for the extraction of wood-boring insect larvae.

The Yellow-tailed Black Cockatoo is the most commonly encountered Black Cockatoo in south-eastern Australia and although in the past a number of subspecies were recognised, currently two subspecies are accepted throughout its extensive range. *Calyptorhynchus funereus funereus* is the nominate subspecies from the northern and central parts of the range and the smaller, shorter tailed *Calyptorhynchus funereus xanthanotus* is from the south. (See Plate No. 6A).

Earliest Reports

This Cockatoo was first described and figured by Shaw and Nodder in "Naturalists Miscellany" Volume 6, Plate 186 in 1794 as *Psittacus funereus* (Funeral Cockatoo).

Plate 5: Yellow-tailed Black Cockatoo *Calyptorhynchus funereus* Female.

The rapidity with which specimens of this species found their way to England and Europe after the settlement of Australia in 1788, was no doubt due to the frequent occurrence of the Yellow-tailed Black Cockatoo in the vicinity of Sydney, where even today it is often seen in the outer suburbs and the adjacent rural areas.

Range Habitat and Field Notes

The Yellow-tailed Black Cockatoo ranges over the coastal, mountain and western slope regions of eastern and south-eastern Australia including Tasmania. It extends from approximately the Drummond Range region of Queensland, south through the eastern third of New South Wales and all over Victoria with the exception of the north-west, Tasmania, the Bass Strait Islands, King Island and the south-eastern coastal region of South Australia from the Gawler area to the Victorian border, plus an isolated population on the southern Eyre Peninsula.

The larger nominate subspecies, *C. f. funereus*, ranges through Queensland, New South Wales and eastern Victoria to an area roughly north of Melbourne where it meets and probably overlaps the range of the smaller *C. f. xanthanotus*, which is found throughout the rest of the range.

This cockatoo leads a nomadic lifestyle, frequenting all types of forested areas, including the introduced pine forests to which so few species of native fauna have been able to adapt, open woodland, coastal scrubs, heathlands and occassionally they inflict crop damage in orchards. Sometimes large flocks of one hundred or more birds may be seen moving noisily through an area gathering food, although sightings of a dozen or so birds or family groups consisting of a pair and their previous season's youngster are more common.

It is usual for young Yellow-tails to fly with their parents until the approach of next year's breeding season, when they are driven away or ignored.

Yellow-tailed Black Cockatoos are reasonably numerous in the Blue Mountains west of Sydney where their approach and presence is usually heralded by their constant calling. Fortunately it is still not an uncommon occurrence to see this cockatoo on the outskirts of Sydney. Looking back over the years it seems that almost every time I have bushwalked in this mountain country I have been rewarded with good sightings of this bird or at least heard their unmistakeable calls in the distance.

Their flight is of a slow ponderous yet powerful nature and probably can be most appreciated when they are observed flying across a vast mountain valley. When this species is flushed in a forested area they show remarkable flight manoeuvreability for such a large bird, as the tree tops are negotiated. Speeds in excess of 60 kilometres per hour have been recorded recently.

Many years ago I observed a small flock of Yellow-tails feeding in low scrub, on the north coast of Tasmania. As I approached they flew a short distance then settled again and each time I tried closer observation the same routine was employed, so as to keep about fifty metres from me. I have found this type of behaviour to be consistent with that of birds on the mainland, when found feeding under similar conditions.

Their food consists of seeds, nuts, fruit and berries from a wide range of native trees and shrubs such as Eucalypts, Banksias, Acacias, Hakea, etc., as well as nectar pollen, blossom and herbage. This cockatoo has proved to be capable of adapting to the changed conditions created by the farming industries and general settlement by supplementing its diet with seeds and nuts of introduced flora such as pines, particularly Radiata pine.

A wide range of insects and their wood boring larvae form a substantial part of the Yellow-tail's diet, with some observers noting a marked increase in their insectiverous tendencies just prior to breeding. Those two keen bird observers and aviculturists from New South Wales, Bill Connor and Russell McAllister, recently observed a small flock of Yellow-tails in the Cessnock area of New South Wales. The birds were feeding on a wood boring larvae known locally as Bondi grubs. The cockatoos were working from the ground and leaving channels, 5 cm or more deep in the trunks of a stand of Eucalypt saplings, as they followed the tunnels made by the grubs. Some of the Eucalypts were more than half chewed through in their efforts to locate the larvae. It appears to be more common for this cockatoo to search for larvae in the trunk and branches well above the ground.

The versatile feeding habits of the Yellow-tail, together with the relatively high numbers throughout its often rugged range, suggests a bright future for this species. There is some concern that the destruction of trees containing suitable nesting sites, in many areas, may be limiting the reproduction of this species. If this is the case, the longevity of this cockatoo (possibly 100 years) would mean that a decline in numbers may not be apparent for decades after breeding has ceased and until it is too late.

From the wonderful work being done by the RAOU and published in their Newsletter No 72, June 1987, titled Yellow-tailed Black Cockatoo Survey progress Report No. 3, I have gleaned the following points.

Most suitable localities throughout their range appear to have small resident populations which possibly consist of a few family groups. These populations seem to confine their movements to areas of approximately 200 kilometres across, in which they follow set feeding patterns. Large sections of the Yellow-tail population gather into loose migratory flocks which consist of numerous small units and family parties with juvenile members. These flocks are often in excess of one hundred birds and at times may number several hundred. They follow established flight paths which often broadly follow major river systems or escarpments. Along the route they feed on known seasonally abundant food sources such as green pine seeds, wood boring grubs, Banksia and Eucalypt blossoms and Hakea seed, etc. Flocks will always gather where the food sources are most extensive. Seeds of the introduced pines and particularly the common *Pinus radiata* have become the most important food source for this cockatoo.

Roosting sites are usually not used for more than a few consecutive nights.

Drinking sites are variable although a preference for a secluded watering place surrounded by bush does emerge. Surveys of nesting sites show a preference for River Red Gums in Victoria and South Australia and Spotted Gums in New South Wales although it is suggested that any type of tree with a suitable hollow may be used. Finally I must applaud the RAOU and the BOC of Australia together with their members for this continuing monumental and invaluable work on the Yellow-tail Black Cockatoo.

Breeding in the Wild

A hole in a tree where the top or a large limb has broken off, often in a Eucalypt, is the usual nesting site. The entrance is often a dead spout entering green timber and the nest is normally near vertical at heights varying from 10 m (32 ft) to 30 m (100 ft) or more above the ground, with an internal diameter of 30 cm (12 in.) to 40 cm (16 in.) and depths of 60 cm (2 ft) to 2 ms (6 ft 6 in.) have been recorded.

The hollow is prepared by the male and the female through chewing wood off the entrance and internal sides, thus developing a porous wood base to the nesting chamber. Breeding normally occurs during the autumn and winter months in the northern parts of the range and graduates to spring and early summer in the central regions and to summer and early autumn in the southern extremities of the range. One white roundish oval egg is usually laid, although often a second egg is laid up to a week after the first. Incubation is carried out by the female, who is fed either on or near the nest site, three or four times daily by the male. This procedure continues for two or three weeks after the eggs hatch, then as the chick grows and is capable of maintaining its own body heat, the female leaves the nest unattended while she feeds. Then as the chick develops further less time is spent with it, until it is fed only morning and evening by both parents.

When two chicks hatch it is usual for the second chick to be neglected and allowed to die, although on rare occassions two chicks are raised. After fledging the young is fed by both parents for several months.

With breeding completed, family groups, small flocks and on occasions large flocks of two hundred or more birds are formed as they follow their semi-nomadic and loosely predictable, seasonally orientated movements in search of food.

These movements are predictable with particular populations of this cockatoo in certain areas; for instance Yellow-tails regularly visit some outlying suburbs of Sydney in the late autumn and they are known to seasonally feed on established stands of pines and in pine forests when the seeds are ripening.

Aviculture

Beyond a few references to hand reared birds kept for pets there are no early or recent avicultural records of this species prior to Bob Lynn's world first breeding. Allan Lendon refers to large numbers of wild caught Yellow-tails appearing on the local market in the late 1950's and suggests that prior to this they were rare in captivity. This accounts for the lack of avicultural literature before the 1960's. There is also evidence that this wild caught stock from South Australia formed the original Yellow-tailed stock in New South Wales aviaries.

The world first breeding of this species occurred in Bob Lynn's aviaries in Sydney, Australia in 1965, and not in 1963 as reported by Alan Lendon in the revised edition of Cayley's "Australian Parrots in Field and Aviary", the information in that book being reprinted from the Parrot Society Magazine.

The following is Bob's account of his early experiences and subsequent breeding of the Yellow-tailed Black Cockatoo for which he was presented with a World First Breeding Award by the Avicultural Society of NSW in 1966.

"I obtained my first Yellow-tails in 1958 from Peter Faye of the Pine Birdfarm, Northmead, and others in 1959 and 1960. Having no idea of their breeding or housing requirements, I eventually constructed a large sheltered aviary to accommodate the most likely pair of Yellow-tails and was prepared for alteration, adaption and experimentation to achieve the ultimate success.

"The new aviary was 12.3 m (40 ft) in length, 3 m (10 ft) wide and 2.7 m (9 ft) high and ran from west to east. It was constructed of galvanised iron pipe frame, covered with 2.5 cm (1 in.) heavy chain wire mesh. I consider heavy gauge 2.5 cm (1 in.) chain wire to be the most suitable wire for Black Cockatoos. The western end was fully sheltered for 3 m (10 ft) and the eastern end or front of the aviary was totally enclosed, including the top for about a third of the full length, with horticultural glass. The remainder of the southern side was enclosed for protection from the prevailing winds. The northern side together with the balance of the top of the aviary was left open. The floor was natural earth which supported the growth of various grasses.

"A natural hollow log about 1.4 m (4 ft 6 in.) high with a hollow fork at the top, providing an ideal entrance to the nesting chamber which was approximately 37 cm (15 in.) in diameter, was located along a creek not far from my home. The log was stood vertically on a concrete base, under shelter and filled to a depth of 60 cm (2 ft) with rotting wood dust.

"The Yellow-tails originally shared their aviary with a pair of Cockatiels.

"A seed mixture of eight parts sunflower, two parts maize one part wheat, two parts oats, one part safflower and a few raw peanuts was the diet that I fed at that time, together with a liberal supply of Banksia pods and pine cones.

"The hen nested for the first time in 1963 but the two eggs were infertile. I wasn't surprised, as the cock appeared to me as if he was unsuitable for breeding purposes and also showed no compatibility with the hen. Fortunately, I had another male in good condition so I replaced the original male for the next season. I estimated that both birds were now over five years old. My experiences later proved that four years old is the usual minimum successful breeding age for this species.

"The next season, 1964, the hen again laid two infertile eggs but she did appear to have accepted her new mate. About this time I commenced feeding a dessertspoon of raw mince meat twice a week and at that time attributed it to my eventual success. Prior to this, I had fed cooked chop bones but at that stage the Yellow-tails had not developed a real liking for them.

"During late December 1964 the hen showed her usual interest in the nesting log and chewed continuously around the entrance hole, dropping much of the woodchips into the log. Banksia nuts and branches from native trees and shrubs were particularly appreciated for chewing materials as breeding time approached.

"On 16 January, 1965, the first egg was laid followed by the second egg on the 18 January and the hen commenced incubation on 19 January. During incubation the cock fed the hen on the nest and she usually left the nest in late afternoon for a drink. The first egg hatched on 19 February after a 30-day incubation period, but the second egg failed to hatch. The chick was covered with yellow down. The hen brooded the chick closely for the first week, then in the second week the chick was left unattended for lengthening periods during the daytime. After 14 days the hen ceased to brood at night but roosted on top of the log. Of course, it must be remembered that this breeding took place during late summer and early autumn when night temperatures are still high, making brooding for a longer period unnecessary.

"On the morning of 17 May I noticed the youngster trying to climb to the mouth of the log and after many attempts it managed to perch just below the entrance. It was still in the same position when I looked again in the afternoon. Ten days later on 17 May the young bird left the nest, 90 days after hatching, but it was weak and unable to fly and had to be lifted to the perch to roost every night. I also found it necessary to supplement the parents' feeding with soft food from a spoon. It was more than a month later, on 19 June, that the youngster took its first flight from the perch to the top of the nesting log. Although the bird managed to fly weakly, it remained delicate, never having reasonable strength and vitality and eventually died at just over 12 months of age.

"During the rearing period the parents were fed sweet corn, canary seed, which the cock bird relished, and mealworms in addition to their normal diet.

"On 15 January, 1966, the same pair laid their first egg and commenced incubation immediately. The egg duly hatched on 14 February after another 30-day incubation period, the second egg failing to hatch. This chick was strong and vigorous from the first day and fledged 90 days after hatching, being capable of strong flight from the first day. Both parents fed the youngster for three months after it left the nest and then it was fed by the male alone until the next breeding season. The pair continued to breed reliably for many years after these first successes."

Sexing

Adult Yellow-tailed Black Cockatoos are readily sexed, the male having a blackish brown beak, pink periophthalmic eye ring and dull yellow ear coverts, whereas the female has a horn coloured beak, dark grey periophthalmic eye ring and bright yellow ear coverts.

Immature birds are similar to the adult female with a grey beak and although it appears that young males often have duller ear coverts and darker grey beaks, while the young females often have brighter ear coverts and lighter coloured beaks, there is no degree of certainty of sex until the birds are 18 months to 24 months of age when they acquire full colour.

Display

I recently observed this species' courtship display in Bob's aviaries. The female was perched on top of the nesting log uttering a high pitched whistling call, whilst the male strutted and bowed on an adjacent perch with his crest up and his tail slightly fanned, making a gutteral clucking sound similar to the clucking of a domestic fowl. The hen is often fed by the male after these courtship displays.

Nests

As previously mentioned, Bob Lynn favours a vertical hollow log about 1.5 m (5 ft) high, with a natural entrance hole and internal diameter of approximately 30-40 cm (12-16 inches). They are placed on the ground under shelter within the aviary but not close to either end of the aviary. The log should be filled to within 60-75 cm (2-2.5 ft) from the top with rotten wood dirt taken from old stumps in the bush. Bob prefers broken up soft tree termite nests for nest log filling and feels that wood chips are too coarse and hard, often causing eggs to be broken.

Mark Schmidt of South Australia stated in his lecture at the 1983 Brisbane National Avicultural Convention that his Yellow-tails preferred a large log hung at approximately thirty-five degrees to vertical.

Nesting and Hatching

This species usually lays one egg but frequently two eggs are laid with a two to eight day interval between them.

During Bob Lynn's early years with this species they constantly nested in his Sydney aviaries in January and occassionally in February, but he now considers that his aviary bred stock is likely to breed at any time of the year. This does not appear to apply to aviary birds in our southern states where summer nesting is normal.

Incubation usually commences with the laying of the first egg, although when two eggs are laid it sometimes starts with the second egg and is carried out by the female alone. Bob has recorded 29 and 30-day incubation periods, whereas Mark Schmidt records a 28-day period in South Australia. Variations of recorded incubation periods in this species or in all psittacines for that matter, apppears to be related to climatic and weather conditions, housing and nesting facilities and how well the female incubates.

The chick hatches covered with a bright yellow down and grows fairly slowly for the first week and then progresses rapidly. The hen does all the feeding of the young bird for the first week or 10 days, then the male starts to assist and in some cases almost takes over the task totally. Bob Lynn has one breeding pair of the northern subspecies *C. f. funereus* in which the male takes no part in feeding the young in the nest although he constantly feeds the hen. The hen broods the chick constantly for the first week or so and then leaves it unattended for increasingly longer periods.

Although Yellow-tails often lay two eggs, usually only one chick is reared. The second chick to hatch, which is often a week behind the first baby is neglected and allowed to die, then buried in the wood dirt at the base of the nest. On some occasions it appears that the second chick is actually murdered as Bob Lynn discovered when investigating the disappearance of the smaller chick in a nest of two. The chick was found to be buried alive in the nesting material. It was in a weakened state and died soon afterwards.

For those breeders who are capable hand rearers, the opportunity to take the second chick as soon as it hatches, or the second egg for artificial incubation, is there to be exploited.

Mark Schmidt, in his informative Brisbane Convention Lecture notes, records a unique breeding in his aviaries during 1981, in which the two chicks hatched were reared to fledging by their parents. He attributed the success to the two eggs having been layed only 48 hours apart, not a week as is often the case. Hence the chicks hatched closer together and the youngest one was able to compete with the oldest one for food.

John Meppen of central New South Wales recently had a nest of two healthy young Yellow-tails reared to 14 days by the parents, but unfortunately they were lost when the hen failed to brood the chicks on an unseasonally cold night.

A Subspecies. Of the two adult male Yellow-tailed Black Cockatoos pictured, the large bird is of the northern race of the nominate subspecies *C. f. funereus* and the small bird is of the southern subspecies *C. f. xanthanotus*.

stling 10 days old.

C Nestling 23 days old.

tling seven weeks old.

E World first parent reared young with parents in background.

Plate 6: Yellow-tailed Black Cockatoos. Subspecies and Nestlings.

At about three weeks of age the eyes begin to open and pin feathers start to appear and the growth rate increases rapidly. Fledging usually occurs between 12 and 13 weeks, when the young leave the nest similar to the adult female with a light grey beak, dull ear coverts and often with wispy traces of the hair-like yellow down.

After fledging young Yellow-tails are often fed by their parents for up to 12 months, with the male accepting most of the responsibility for this duty. Young birds may be safely removed from their parents four months after leaving the nest.

Bob feeds large quantities of Hakea and Banksia nuts as supplements to the normal diet, during the entire rearing period. Young aviary bred Yellow-tails have been known to lay at three years but it is doubtful if young males are fertile at the same age.

This species has always been a nervous breeder in Bob's aviaries and he prefers to house breeding pairs in secluded aviaries which receive no disturbances.

Mutations and Hybrids

Birds with extensive yellow suffusion have been recorded on several occasions in the field. Several recent sightings of "yellow" Yellow-tailed Black Cockatoos have occurred in Victoria and on Kangaroo Island. Published sightings at Linton, Sale and Warnambool in Victoria are at hand. The following eye witness report by that Victorian father of aviculture Ron Hastings is invaluable. Early in 1987 Ron first observed this yellow variant feeding in a stand of pines on his farm at Bacchus Marsh. The bird appeared basically yellow with dark feathers in the tail and blackish blotches on the back. At the same time and locality in 1988 a yellow bird appeared again (possibly the same bird) with black tips to the flight feathers, less dark blotches on the back and one dark tail feather in its yellow tail. Perhaps this was a mutant but more probably a colour variant. On the same day that Ron observed this bird in early 1988 a friend of his from near Warnambool over 100 kilometres away, saw another yellow bird within an hour or two of Ron's sighting, so obviously there are at least two yellow Yellow-tailed Black Cockatoos in Victoria at the moment.

I can find no record of hybrids occurring with other species.

WHITE-TAILED BLACK COCKATOO

Plate 7: White-tailed Black Cockatoo *Calyptorhynchus baudinii* — Male.

WHITE-TAILED BLACK COCKATOO

(Calyptorhynchus baudinii)

Derivation:

Calyptorhynchus:
calyptos — Greek for hidden
rhynchus — Greek for beak
baudinii — After Nicolas Baudin who led a French scientific expedition to Australia in 1802-03

Description:

Length 65 cm
Weight 700 g average
Appearance as in Plates Nos. 7 and 8

Classification

Two subspecies are recognised, *Calyptorhynchus baudinii baudinii*, which has a longer, more slender bill, adapted for the gathering of its main food sources, such as Eucalypt seeds (marri and karri) and wood boring insect larvae. (See Plate 9A.) *Calyptorhynchus baudinii latirostris* has a shorter broader bill and inhabits the drier regions of south-west Western Australia. (See Plate 9B.)

Both subspecies have distinct breeding ranges with *C. b. baudinii* being confined to the densely forested south-western corner of Western Australia, whereas *C. b. latirostris* breeds in the drier, more open forest country. During the non-breeding season their nomadic food gathering causes their range to overlap in some areas.

As an aviculturist I cannot accept the current trend in ornithological literature to classify the White-tailed Black Cockatoo as a subspecies of the Yellow-tailed Black Cockatoo. They are obviously closely allied species, but in my opinion the complete isolation of one from the other for thousands of years has allowed two separate species to evolve.

I find it interesting that some taxonomists and ornithologists now consider the nominate subspecies *C. b. baudinii* to be a distinct and separate species and *C. b. latirostris* to be a subspecies of *C. f. funereus*, the Yellow-tailed Black Cockatoo. This theory is difficult for me to accept, for I

Plate 8: White-tailed Black Cockatoo *Calyptorhynchus baudinii* — Female.

cannot see how two supposedly distinct species could possibly evolve to be almost identical birds except for variation in size and shape of the mandibles, which have obviously developed due to the different feeding habits necessitated by the habitat variations of their distinct ranges.

Other differences cited between *C. b. baudinii* and *C. b. latirostris* such as diet, length of wing, size and call variations are all present in the subspecies of Galahs, Corellas and Sulphur-crested Cockatoos yet there is no suggestion of forming new species in these groups, in fact the trend is the exact opposite.

Earliest Report

Although this species derived its scientific name from the French scientist, Nicolas Baudin, who led an expedition to Australia in 1802, it was either not encountered by the expedition or, if encountered, not recorded by the expedition, for the first published record of this cockatoo was not until Lear's illustration in his "Illustrations of the Psittacines" plate 6, in 1832. The first recorded natural history of this species was by Gould in his "Birds of Australia" Volume 5 plate 13, in 1846.

Range, Habitat and Field Notes

The White-tailed Black Cockatoo ranges over the south-west region of Western Australia from the Murchison River in the north, inland to the Lake Moore region and down to Norseman, then east to Israelite Bay. The nominate subspecies *C. b. baudinii* inhabits the densely forested areas of the extreme south-west although occasionally wanders north beyond the Jarrah and Karri forests and has been observed feeding in the company of *C. b. latirostris* which inhabits the more extensive dryer areas of the range.

Some years ago, during a visit to Western Australia, I was able to gain odd glimpses of pairs of *C. b. baudinii* in the beautiful tall forests of the south-west. I found it more difficult to observe than *C. b. latirostris* mainly due to the type of country it inhabits.

Like the Yellow-tailed, this cockatoo's presence in an area is usually made known by its loud calls which appear to differ from the Yellow-tail in being deeper in tone and less musical. Saunders has noted a difference in the contact call of the two subspecies of the White-tailed Black Cockatoo.

Comparatively little movement is recorded within the range of *C. b. baudinii* and in some areas it is believed to be sedentary, whereas constant seasonal movements have been recorded throughout the more arid range of *C. b. latirostris*.

I was able to observe good numbers of *C. b. latirostris* during early September just north of Perth, while numerous other sightings were made within 100 kilometres north and east of Perth. While setting up camp at dusk in Yanchep National Park, several White-tails were seen flying over. Then soon after dawn the following morning, many small flocks passed over, having left their roosting sites in the park to fly to the feeding grounds. We were leaving the National Park, when a kilometre or two from the camping area a flock exceeding 50 birds was sighted feeding in low scrub, mainly on the seed pods of a small leafed Banksia.

Estimated populations of both subspecies of this cockatoo are quite high, but there is much concern for their long-term well-being due to the reduced breeding activities in many areas, probably caused by the clearing of feeding grounds in the vicinity of suitable nesting sites and vice versa.

Like the Yellow-tail this species forages in family parties, consisting of the breeding pair and their young, as well as small groups, and apparently large flocks are often formed, particularly in *C. b. latirostris*, soon after breeding. These large flocks migrate to the higher rainfall areas nearer the coast, in search of food, where they forage nomadically until the approach of the breeding season, when they again return to the dryer inland regions to breed.

In the far south-west *C. b. baudinii* feeds mainly on Eucalypt seeds, particularly Marri and the larvae of wood boring insects. Crop damage in apple orchards has been recorded from this subspecies.

C. b. latirostris, from the more extensive drier region of the south-west, feed on the seeds of native trees and shrubs, particularly Eucalypts, Dryandra, Hakea, various introduced pines and the seeds of many grasses and ground plants. This species is often seen feeding on the ground from fallen seed pods and the seed of ground flora, unlike the Yellow-tail who seldom comes to the ground except to drink.

It is interesting to note that *C. b. latirostris* has developed a liking for Pinus seeds yet *C. b. baudinii* has not. The White-tailed Black Cockatoo flies similarly to the Yellow-tailed Black Cockatoo with a slow strong wing beat and flies very high when moving over long distances. This cockatoo is also capable of fast erratic flight through forest tree tops when aroused.

Breeding in the Wild

The nesting site is usually in a large hollow limb or hole in the trunk of a large Eucalypt tree, which may vary in depth from 50 cm (20 in.) to

2 m (6 ft 6 in.) or more. Both male and female commence to prepare the nesting chamber many weeks before the first egg is laid. The male often displays and feeds the female during this period of wood chewing in the hollow.

A clutch of one or two white rounded oval eggs are laid on a bed of wood chips and splinters, often as early as July and all nesting activities are completed by the end of December. When a two-egg clutch is laid there may be up to a week interval between eggs. Incubation is carried out by the female, who is called off the nest and fed by the male, but she quickly returns to her duties, entering the hollow, rear first. This feeding procedure is continued for two or three weeks after the egg hatches, then as the chick grows, brooding and feeding progressively decreases, until it is left unattended during the day, being fed only morning and evening by both parents.

Fledging occurs about three months after hatching and the youngster is fed for a further four months by the parents. When breeding is completed they form family parties and flocks and in the case of *C. b. latirostris* migrate to coastal areas.

Usually when a two-egg clutch is laid, the female ceases to incubate the second egg once the first egg hatches which results in the death of the embryo in the second egg. On the odd occasions that two eggs do hatch, the youngest is usually neglected and allowed to die. Rarely has the successful rearing of two young been recorded.

Aviculture

I can find no references to this species in early avicultural literature. There was a pair held in Adelaide Zoo during the mid 1930's. Cayley states that this was the first pair he had seen in captivity, which indicates its rarity in aviculture at that time.

Alan Lendon records having seen quite a number in Sir Edward Hallstrom's Sydney aviaries in June 1949. At that time eggs had been laid on a couple of occasions. Lendon also states: "Even R. T. Lynn of the Sydney suburb of Condell Park does not appear to have succeded yet, although he had two fine pair when I saw his collection in October 1969." In November 1970, Graham Taylor, then of Terry Hills in Sydney, had a pair lay one egg, which they duly hatched, but the chick only survived until six weeks old.

Forshaw states that Saunders (1976) recorded the successful breeding of the White-tailed Black Cockatoo at the CSIRO Division of Wildlife Research Laboratories near Perth.

The White-tailed Black Cockatoo was the last of the five species of Black Cockatoos to be bred by Bob Lynn. Bob considers this was not because they were more difficult to breed but that his pairs were slow to co-operate. Bob obtained his first White-tails in 1961 and he first bred this cockatoo in 1974. At the time he did not consider this breeding to be a world first as he had been told the species had been previously bred in Western Australia. After exhaustive inquiries and extensive research into the avicultural literature, I am convinced that this breeding is also a world first or at the very least the first recorded world breeding of the White-tailed Black Cockatoo. The single egg was laid on 21 October, 1973, and duly hatched after a 30-day incubation period on 20 November, 1973. A basic diet of sunflower seed supplemented with corn, oats, safflower, green food and plenty of the White-tails' favorite native food, Hakea nuts, was supplied. Hakea nuts are extremely hard seed pods about the size of a small walnut and contain a single tiny seed. The youngster left the nest on 15 February, 1974 after 87 days, with its legs slightly affected by rickets, a condition which required treatment with a calcium supplement.

Rickets, or Metabolic bone disease, was a recurring problem with Bob's young White-tails until he was able to introduce the pairs to Banksia cones. Once they commenced eating the Banksia seeds and in turn fed them to their young, the improvement in the quality of the babies reared was unbelievable. Obviously the nutrients contained in Hakea is insufficient to rear robust youngsters, although Hakea is their preferred native food. Banksia cones are now relished by all of Bob's White-tails.

Many other successful breedings of this species have occurred in Bob's aviaries since the 1974 success, including another world first achievement, when in 1986, two fine young specimens were fledged from the one nest. Bob's own version of this unique breeding is as follows:

"The pair are housed in an aviary approximately 12 m (40 ft) long, 3 m (10 ft) wide and 2.7 m (9 ft) high, with a similar aspect and shelter as the aviary in which I first bred the Yellow-tails, in fact it is adjacent to this aviary. The nesting log is similar to those used for all my Black Cockatoos except the entrance is slightly smaller and this pair nests deeper (almost 90 cm or 3 ft) than most of my cockatoos, a condition which made Stan's photography of these young more difficult, due to the smaller hole to photograph through and the darkness inside the log making focusing difficult. The first egg was laid on 20 October, 1985, and incubation commenced immediatedly, then the second egg was laid on 27 October, 1985 — a weeks interval which normally eliminates any chance of success with the second egg. The first egg hatched on 18 November, 1985, after a

29-day incubation period and the second egg hatched on 24 November, 1985, after an incubation period of 28 days. I found the six-day difference in age to be obvious in the nestlings during the entire growing period.

"On 6 February, 1986, the first chick to hatch left the nest after 80 days and was an obvious male, then on 13 February, 1986, the second youngster flew after 81 days in the nest and it was a very obvious hen.

"This breeding was not only unusual because two young were reared in the one nest but also because the sex of the birds could be easily determined by the brightness and extent of the ear coverts and the colour of the beak. Of all the young White-tails I have seen and bred, these are the only youngsters I have been able to sex with surety. I have been able to sex some at two years old, yet others have been three or even four years old before I was sure."

Bob left these two youngsters with their parents for almost 12 months, but believes that in doing so he delayed the next season's nesting for a month. Ten days after the young were removed from the aviary, the hen laid on 24 November, 1986, and a second egg was laid on 30 November, 1986. The first egg was infertile, possibly due to its being laid so soon after the previous young were removed, but the second egg hatched after a 30-day incubation on 30 December, 1986, and fledged 83 days later on 23 March, 1987.

Since Bob's early successes with White-tails their diet has been expanded to include sprouted corn (maize) and sprouted sunflower seed and of course the Banksia pods as well as the items already mentioned.

The first recorded breeding of the Long-billed White-tailed Black Cockatoo, *C. b. baudinii*, appears to be that achieved by Fred Bohner in 1982. He was then a resident of South Australia and was awarded the R. W. McKechnie Memorial Medal by the Avicultural Society of South Australia for this first breeding. The following is an abbreviation of the account of the notable breeding which was published in "Bird Keeping in Australia" Volume 27 No. 2, February 1984.

The pair were housed in an aviary 8 m (26 ft) long, 2.4 m (8 ft) high and 1.5 m (5 ft) wide. Construction was with 25 mm (1 in.) square steel tubing with 25 mm (1 in.) square heavy gauge weld mesh wire. Two-thirds of the top of the aviary was roofed in and some fibreglass sheeting illuminated the sheltered area. A large log stood on the floor of the aviary to provide the nesting facility. This pair of White-tails were in Fred's collection for six years before any breeding attempt was made. Then in November 1979, two eggs were laid, which proved to be infertile, followed by two more infertile eggs in January 1980. The next season one egg was fertile but the chick died during the early stage of incubation.

The succesful breeding commenced with an egg laid on 5 November, 1981, followed by a second egg on 7 November. Incubation started with the laying of the second egg and was carried out by the female. Prior to the hen laying the male was seen to drop several pine and Banksia cones into the log, which were chewed up by the female to provide added nesting material.

Two chicks hatched on 4 December, 1981, but one died on the second day and was not removed from the log. The incubation period was exactly 28 days and the chicks hatched with a pale yellowish down. Daytime temperatures of 40°C for a week's duration seemed to lessen the chick's chance of survival, but added ventilation was provided by opening an inspection door in the log and this probably saved the chick's life. After 16 weeks the youngster left the nest, although unable to fly due to a damaged wing, which obviously extended the length of time spent in the nest.

A basic seed diet of sunflower seed with a little oats and canary seed was fed, although little of the latter seeds were eaten. Pine cones, Banksia cones, *Eucalypt ficifolia* seed pods, almonds in the shell and green panic grass stems were all fed as supplementary foods.

The following observations and breeding data of the Long-billed White-tailed Black Cockatoo *C. b. baudinii* were recorded in Perth Zoo by Neil Hamilton and Peter Brindley, then kindly made available to us by John De Jose, the Director.

Unusual drawings in the sand on the floor of the aviary housing a mated pair of this cockatoo have been found to be executed by the male with his upper mandible during what is assumed to be an unusual form of courtship display directed towards the female prior to and during the breeding season. These drawings are variable and often quite extensive, sometimes covering the entire sanded floor area which is 6 m (19 ft 6 in.) long and 2.1 m (6 ft 10 in.) wide.

I can find no other records of similar behaviour from this subspecies of cockatoo. Perhaps further observations will determine if these artistic skills are in fact related to courtship display or just an individual talent on the part of this male. This pair of cockatoos have been in the zoo since 1983 and are housed in an aviary constructed of 25 mm (1 in.) galvanised pipe and 75 mm (3 in.) by 25 mm (1 in.) wire mesh. The aviary is 9 m (29 ft) long by 2.1 m (6 ft 10 in.) wide and 2.4 m (7 ft 10 in.) high, with the

A Nominate subspecies *C. b. baudinii.* Note long slender beak. Female.

B Subspecies *C. b. latirostris* with shorter, broader beak.

Plate 9: White-tailed Black Cockatoo — Subspecies.

C Nestling nine days old.

D Nestlings seven weeks old. Note: Previousl
unrecorded, two chicks reared in the one n

E Nestlings 10 weeks old.

F Nestlings 11 weeks old just prior to fledging.

Plate 9: White-tailed Black Cockatoo — Nestlings.

rear 2 m (6 ft 6 in.) roofed and sheltered. The front 6 m of the floor are covered by sand and the remaining 3 m by concrete. Perches of sawn hardwood and gum branches are provided at each end of the aviary.

The nesting log is 1.2 m (4 ft) long with an internal diameter of 35 cm (14 in.), the bottom is sealed with perforated galvanised iron sheet and the entrance is down the top of the log. It is stood vertically on a 65 cm (2 ft 2 in.) metal stand, on the left-hand side of the open section of the aviary, about 5 m from the front. Flat iron is placed on top of the aviary above the nest to protect it from adverse weather conditions. Natural wood dirt 15 cm (6 in.) deep is placed in the bottom of the log.

As early as July in 1987 the pair were seen chewing around the log although the hen was not seen in the log until the middle of November. The pair were seen copulating on the 3 December, one egg was laid on 6 December but broken the next day. Another egg was laid on 16 January, 1988 and incubation by the female commenced the next day and was continued diligently until 12 February when the hen became unsettled and was disturbed by the slightest noise. The egg was removed on the 20th and found to contain a dead but well developed chick, which it was thought may have died from excessive heat. Within six weeks the female was working the log again and the male feeding her. Hopefully success will follow.

Sexing

Adult White-tailed Black Cockatoos are easily sexed with the male having a dark grey beak, pink periophthalmic eye ring and dirty white ear coverts. The female has a horn-coloured beak, dark grey periophthalmic eye ring and clear white ear coverts. Immature birds are similar to the adult female with grey suffused beaks and dull white ear coverts. With the odd exception young White-tails often remain difficult to sex until three or four years of age.

Display

The display is similar to the Yellow-tail, with the male approaching the female in a strutting movement, his tail fanned, his crest erected and uttering a low musical clucking sound, as he bows periodically. The male often feeds the female after his display.

Nests

Bob Lynn favours his usual vertical hollow log about 1.5 m (5 ft) high, with some form of natural looking entrance and an internal diameter of approximately 30 cm (12 in.) to 40 cm (16 in.), standing on the ground, under shelter and away from either end of the aviary. The log should be filled to within 60 cm (2 ft) to 75 cm (2 ft 6 in.) of the entrance with wood dirt or preferably crushed, soft type termite nest.

Nesting and Hatching

Both male and female prepare the nesting log prior to laying one or often two white oval eggs at an interval of between two and seven days. Bob Lynn's White-tails lay two eggs more often than one. Incubation is carried out by the female, who is fed on or near the nest by the male.

We have recorded fresh egg weight of 33.2 g.

Incubation periods of 28, 29 and 30 days have been recorded in Bob's aviaries. Chicks have hatched with down varying in colour from yellow to off-white. Bob feels that down colour variations could be related to diet or perhaps weather conditions. When two eggs are hatched the youngest chick is usually neglected and allowed to die.

Bob has observed that in his aviaries White-tail chicks are fed in the nest by both male and female. The eyes start to open and pin feathers appear from two to three weeks of age; no secondary down has been observed. Fledging usually occurs between 80 and 90 days after hatching when the youngsters appear similar to adult females with pale grey beaks, dull ear coverts and often with hair-like traces of yellow down. Both parents continue to feed the young for three to four months after fledging, at which age Bob feels it safe to remove the young from their parents. White-tail Black Cockatoos are sexually mature at four years of age.

All nesting activities in Bob Lynn's aviaries have commenced with eggs being laid in October or November, which does not closely conform with recorded field nesting operations, when eggs have been noted from July to October. Bob also records that White-tails like to bath, unlike other Black Cockatoos.

Mutations and Hybrids

I have been unable to find record of mutations occurring in this species or of hybrids occurring with any other species.

RED-TAILED BLACK COCKATOO

Plate 10: Red-tailed Black Cockatoo *Calyptorhynchus magnificus* — Male.

RED-TAILED BLACK COCKATOO

(Calyptorhynchus magnificus)

Derivation:

Calyptorhynchus:
calyptos — Greek for hidden
rhynchus — Greek for beak
magnificus — Latin for magnificent

Description:

Length 65 cm average
Weight 730 g average
Appearance as in Plates Nos. 10 and 11

Classification

As with all members of this genus, the particular development of the mandibles reflects the degree of specialisation of their feeding mode. The less specialised feeding habits of *Calyptorhynchus magnificus* have developed an intermediate type mandible, less suited to excavating wood boring grubs than the long, pointed, narrow mandibles of *Calyptorhynchus funereus* and *Calyptorhynchus baudinii*. However, the medium pointed upper mandible and the chisel edged lower mandible is far more suited to cracking open a wide range of nuts, fruits and seed pods than the specialised heavy duty mandibles of *Calyptorhynchus lathami*.

The research carried out for the RAOU's "Atlas of Australian Birds" presents evidence of at least seven and possibly eight distinct populations, whose relationships have not as yet been fully established. Obviously until extensive field work is carried out on this species, the correct number of subspecies and the differentiation between them will remain unknown.

Currently four subspecies are generally recognised as sugggested by Forshaw in his revised second edition of "Australian Parrots".

Calyptorhynchus magnificus magnificus is the nominate subspecies and is a noticeably large subspecies in which the female has yellow barring on the tail, progressively graduating to orange near the tip. (See Plate No. 12A.)

Calyptorhynchus magnificus macrorhynchus is also a large subspecies which has a broader, heavier built beak and the female has pale yellow

Plate 11: Red-tailed Black Cockatoo *Calyptorhynchus magnificus* — Female.

barring on the tail, with some specimens showing a slight trace of orange near the tip. (See Plate No. 12B.)

The range of *C. m. magnificus* meets and possibly merges with that of *C. m. macrorhynchus* in the Gulf of Carpentaria region.

Calyptorhynchus magnificus samueli is a smaller subspecies with a smaller, narrower beak. It has a predominately inland range with four distinct isolated populations. (See Plate No. 12C.)

Calyptorhynchus magnificus naso is also a small subspecies similar in size to samueli, but with a broader beak. This subspecies has a shorter more rounded crest in comparison to the other subspecies. There are two distinct populations, one from the heavily forested areas of the south-west of Western Australia and the other from the drier open Eucalypt forests of Western Australia. (See Plate No. 12D.)

Earliest Reports

Unlike many of our Cockatoos, this species received considerable attention from the early explorers and naturalists who visited the east coast of Australia. This suggests that the Red-tailed Black Cockatoo inhabited the Botany Bay region of Sydney 200 years ago.

The first illustration was a pencil sketch of a female drawn in 1770 by Sydney Parkinson, who sailed in the *Endeavour* under Captain James Cook and was an illustrator draughtsman assisting Joseph Banks. This sketch is not only the earliest known of an Australian cockatoo, but the first of any Eastern Australian bird.

Latham, in his "General Synopsis of Birds" Volume 1, page 260, 1781, refers to Black Cockatoos (Red-tailed) in a generalised way and relates to Parkinson's Voyage on page 144. Then in his "Supplement to a General Synopsis of Birds" 1787, page 63, Latham described the Red-tailed Black Cockatoo and figured it in Plate 109, but failed to provide it with a scientific name. Later again in 1790 in his "Index to Ornithology" Volume 1 page 107 he provided this species with the Latin name of *Psittacus banksii*.

Also in 1790, Shaw and Nodder described and figured this species in their "Naturalists Miscellany" Volume 2 plate 50 as *Psittacus magnificus*. This name eventually gained preference over Latham's *Psittacus banksii*, although this species is still commonly referred to as the Banksian Cockatoo.

A Nominate subspecies *C. m. magnificus*. Note long, full crest.

B *C. m. macrorhynchus*. Note broad heavily beak and yellow barred tail of female.

C *C. m. samueli* is a smaller subspecies. Note intermediate length crest, and small mandibles.

D *C. m. naso* is a small subspecies. Note the s full, rounded crest.

Plate 12 : Red-tailed Black Cockatoo — Subspecies.

Range, Habitat and Field Notes

The Red-tailed Black Cockatoo ranges over much of the Australian continent but is absent from Tasmania. It inhabits the whole of northern and north-eastern Australia, together with several isolated populations in central, western and southern regions.

The exact range of the accepted subspecies remains undetermined hence the following locality data is only as accurate as the information currently available.

C. m. magnificus, the nominate subspecies, ranges over the entirety of north-eastern Australia from somewhere in the Gulf of Carpentaria, where it merges with *C. m. macrorhynchus* and extends inland along the Gulf river systems to approximately the Camooweal area of Queensland, then eastward, taking in all of Cape York Peninsula and south-eastward to north-eastern New South Wales. This subspecies probably gives way to *C. m. samueli* where it encounters the dryer plain country of central Queensland.

C. m. macrorhynchus ranges over the northern region of Australia from where it meets *C. m. magnificus* in the Gulf of Carpentaria, westward through the northern half of the Northern Territory to the vicinity of Broome in Western Australia and south to the Sturt Creek area.

C. m. samueli has a number of apparently isolated populations, mainly in arid inland areas, which may even prove to be all separate subspecies, although this is unlikely when considering the current "diminishing species" trend in ornithological classification. There is one population in the southern part of the Northern Territory which broadly speaking inhabits the Macdonnell Ranges, adjacent high country and river systems, being confined by the surrounding deserts and arid plain country. Another population is situated east of the previous group and isolated from it by the Simpson Desert. The range extends along the Diamantina River system of south-west Queensland and the Coopers Creek—Lake Eyre region of north-eastern South Australia.

A third population inhabits the areas adjacent to the Darling River system and its northern tributaries from the New South Wales-Queensland border in the north to the junction of the Darling and Murray Rivers in the south. Yet a fourth population exists in the south-east corner of South Australia and the south-western corner of Victoria, where the range seems to extend from Bordertown in the north to Mount Gambier and Casterton in the south. Mark Schmidt, of South Australia, who has extensive field experience with this population, considers them to be the smallest of the

Red-tailed group, although the general concensus is that *C. m. naso* is the smallest. I have not had the opportunity to compare captive specimens.

It has been suggested that the small isolated populations along the DeGrey and Oakover River systems on the edge of the Great Sandy Desert may be more closely allied to *C. m. samueli* than to *C. m. naso*.

C. m. naso has two distinct populations — one from heavily forested regions of the extreme south-west corner of Western Australia, extending eastward to Albany and separated from the northern population by a 50-kilometre gap situated just south of Perth. The northern population of *C. m. naso* extends from around Perth, along the coastal regions to a little north of the Murchison River, eastward to the Parker Range, Lake Ballard and Lake Austin, then extending northward from the Murchison River region, inland to the Gascoyne River.

It is interesting to note that the southern population of *C. m. naso* from the Marri and Jarrah forests have developed a longer beak to cope with their arboreal food sources, as opposed to the shorter, broader beak of the northern population which is more suited to feeding in dryer Ecalypt forest and coastal scrubs, as well as feeding on the ground. This distinctive beak development of the two populations is parallel to the beak variation in the two populations of the White-tailed Black Cockatoo *(Calyptorhynchus baudinii)* which broadly inhabits the same regions.

I have observed *C. m. magnificus* many times on my travels through Queensland, but my most exciting experience with this subspecies was in June 1984 when on a trip to the tip of Cape York Peninsula. We made camp on the Laura River in early afternoon. Many Red-tails were seen feeding in the surrounding, dry open Eucalypt forest on Eucalypt seed, as well as foraging on the ground for grass seeds in the company of Sulphur-crested Cockatoos. The black and white checkerboard effect on the dry grassland was a unique sight and when the intermingled flocks were flushed, the spectacle was superb as both species flew noisily away to perch in surrounding trees.

As the afternoon wore on it became apparent to us that we had made camp close to the roosting site of these Black Cockatoos. Small parties were constantly arriving along the river after their day's feeding in the surrounding open forest. It was obvious from the incessant calling of these cockatoos that their favourite watering hole and roosting site was no more than 500 m from our camp. Closer investigation showed many hundreds of Red-tails gathered in the area and small groups flying to and from the bank of a broad, clear section of the river to drink. All the cockatoos appeared to roost in the neighbourhood, with as many as 20 birds being seen to settle down for the night in the same tree.

Next morning before sunrise the constant excited calling commenced again, then, as the sun rose, small parties moved up and down the river. An hour or two after sunrise small flocks fanned out in all directions to feed in the surrounding open forest, but a few flocks remained in the vicinity of the river all day. I did not observe these birds to cease their feeding activity in the heat of the day, when the temperature reached the low 30 degrees Celsius. This is unlike most Australian psittacines who rest during the heat of the day. Perhaps it being the coolest time of the year had some significance. The number of Red-tailed Black Cockatoos congregated in this area was estimated to be in excess of one thousand. On our return journey two weeks later the Red-tails were still camping and drinking in the same area.

On my first two trips to the Northern Territory I found *C. m. macrorhynchus* to be by far the most commonly seen cockatoo in the northern region. During my second trip to the Territory in July 1986, they were observed in good flocks all through Kakadu National Park, Arnhem Land and on the Coburg Peninsula, as well as on the outskirts of Darwin. While on Coburg Peninsula I had the unusual sighting of a pair of this subspecies flying well out to sea, across the entrance to Port Essington. where John Gilbert, who was Gould's collector, found them so common 150 years earlier.

When we called at the mining town of Jabiru, on the edge of Kakadu National Park, to purchase more provisions, I was amazed to see these cockatoos feeding in the car park of a modern shopping complex, where dozens of cars were constantly coming and going.The food source was the seeds of stunted Eucalypts, known locally as Bloodwoods. The Red-tails seemed indifferent to all the activity and allowed my close observation. Although I had many close field sightings on this trip, I found it difficult to detect the larger broader bill which differentiates this subspecies from *C. m. magnificus*, although the lack of orange barring in the tail of the females was readily observed.

My notes of this trip suggest that I found the calls of *C. m. macrorhynchus* to be less musical than the other subspecies and similar to some notes of the Corella and Sulphur-crested Cockatoo.

While in Kakadu National Park I photographed a pair working a hollow and chewing the bark from the entrance spout which was only 6 m (20 ft) from the ground in a Eucalypt, standing in tree studded grassland.

My most recent and memorable encounter with this subspecies was in December 1987 whilst a guest on a well known cattle station in the Northern Territory. At our host's request we were taken bird watching in one of the station's helicopters. The pilot was well versed in the local wildlife and their locations, obviously having piloted numerous flights to show his passengers the abundant wildlife on the two and a half million acre property. After flying towards the sea and the mouth of the Daley River for 20 or so kilometres over dry open Eucalypt forest, odd patches of monsoon forest started to appear.

Our pilot swooped low over a particular patch of this forest and flushed a flock of over two hundred Torres Strait Pigeons, which circled around in a manner reminiscent of homing pigeons. The next large patch of forest held 30 or 40 *C. m. macrorhynchus* feeding in the trees. As the helicopter approached they fled the trees, squawking, twisting and turning, trying to avoid what must have appeared to them to be a giant predator. The helicopter followed in hot pursuit, imitating every acrobatic feat the cockatoos employed in their effort to escape. At times the helicopter was flying on its side, following individuals in tight circles with the tree tops spinning uncomfortably close below us. We were all so involved in the chase that it did not occur to us to be frightened, although my body was reacting violently from motion sickness. Then the pilots voice came through the headsets: "Watch them flash their red tails — it must be some form of defence mechanism."

What he was referring to was that when these cockatoos are in normal flight and viewed from slightly above, the red tail feathers are not visible, but when the helicopter came too close to the birds and they felt threatened, the males would fan their tails thus flashing their red feathers at their tormentor. The exact motive for this flashing red tail display can only be supposition, but as the male Red-tail fans his tail to show its red feathers to advantage during courtship display, it is reasonable to assume that the flashing red tail feathers during flight while under stress is related to an alarm or threat display.

Eventually the flock dispersed in every direction after what seemed like ages but could have been no more than a minute or two and we flew on to see the vast flocks of waterfowl on the flood plains — but that is another story.

I have seen *C. m. samueli* several times throughout inland Australia. My first sighting was of a pair feeding in a Eucalypt tree near the banks of the Darling River near Bourke in western New South Wales. Some years later I observed a flock of about 50 birds feeding on the ground, adjacent to a dry creek bed near Winton in central Queensland. It is possible that *C. m. samueli*'s range may extend much further than currently supposed, particularly in good seasons.

During my only trip to Western Australia I observed *C. m. naso* on three occasions, firstly a fleeting glance of a pair on the edge of a karri forest near Pemberton. If my memory serves me correctly, they would have been of the larger billed race from that locality. My next sighting was of a small flock feeding in low eucalypt forest north of Perth. The third sighting was on the same day in low scrub country in the Morawa region when I saw a small flock of Red-tails feeding in the low trees as well as on fallen seed pods on the ground. On each occasion I have seen Red-tailed Black Cockatoos in the field, regardless of the subspecies the sighting has been associated with much loud calling. Some of the notes could well be termed musical, or perhaps this is only to my ears, as I must confess that I find the call notes of all Australian psittacines in the field music to my ears.

Red-tailed Black Cockatoos feed mainly on Eucalypt seeds, such as the Bloodwoods and Woollybut of northern Australia, the Marri of south-western Australia and Stringybarks of southern Australia, together with various native fruits, nectar, blossom, insect larvae and the seeds of various other native trees and shrubs such as Banksia, Casuarina, Hakea, Grevillea and Acacia. In Western Australia the northern race of *C. m. naso* has adapted to feeding on the seed of the introduced double-gee and in some areas this food source comprises in excess of 90% of its total diet. The inland subspecies *C. m. samueli* is more adapted to a ground-based food source such as grass seeds and the seeds of various native and introduced ground flora. The southern isolated population is the exception and it feeds mainly on Eucalypt seed.

This species' flight is strong with long pronouced wing beats. They are capable of flying great distances often at a considerable height. The movements of this cockatoo are usually nomadic and related to the availability of food sources. In some areas its appearance is seasonal and dependent on the ripening of a preferred food item or the availability of nesting sites.

Breeding in the Wild

The preferred nesting site is usually a large hollow limb or hole in the trunk of a Eucalypt tree or a dead tree, mostly in excess of 15 m (49 ft) above the ground, although occasional low nests have been recorded.

The nesting chamber may vary in depths of up to 2 m (6 ft 6 in.) and is prepared by the male and female chewing around the inside of the hollow to form a soft base of chewed and decayed wood for the nest. During the preparation of the nest site the male can often be observed displaying to the female.

Breeding occurs at varying times throughout the extensive range, but basically appears to take place during autumn and winter in the northern regions and the late winter and spring in the central and southern parts of the range. Double brooding has been recorded in south-western Australia with *C. m. naso*, the first nesting occurring from July to October and the second nesting commencing in March or April. The isolated population of *C. m. samueli* from the southern areas adjacent to the Victorian-South Australian border is also suspected of double brooding, nesting having been recorded during October-November and again in February-March.

One oval white egg (very rarely two) is laid and incubation is usually commenced immediately by the female only. She is fed by the male, either near to or at the entrance of the nest in the morning and late afternoon. The chick hatches after approximately a 30-day incubation period and the female continues to brood the chick closely for three weeks, whilst being fed by the male. She then ceases to brood the chick during the day and spends much of her time feeding, the chick being fed by both the male and female during mid morning and again in the late afternoon. When the chick develops pin feathers, night brooding is also discontinued except during inclement weather.

Fledging occurs about three months after hatching and the youngster is fed by both parents for a further three to four months. After breeding, small and sometimes large flocks are gradually formed as the family parties wander in search of food.

Aviculture

As with other members of this genus, early avicultural references are non-existent. There are odd references made in ornithological literature to hand-reared specimens being kept as pets.

The first partial breeding success and, in fact, the first avicultural reference I can find, occurred in the Marquis of Tavistock's (Duke of Bedford) aviaries in England. An egg laid and incubated on the aviary floor was hatched on 26 December, 1939, after an obviously incorrectly recorded incubation period of two months. At almost five weeks of age the youngster developed rickets and was taken by the keeper, Mr Gorman, and was successfully hand reared. Due to the period of hand rearing I believe this breeding does not qualify as a world first. (Avicultural Magazine, 1940 page 136).

Prestwich in his "Records of Parrots Bred in Captivity" part two records that Alan Lendon of South Australia, *in litt.* 26 October, 1944, states that this species was successfully bred by E. Hallstrom (later Sir Edward Hallstrom) of Sydney, New South Wales, in 1943. Apart from an incubation period of 29 days no other data appears to have been recorded. Hallstrom, 8 December, 1948, writes "I am breeding regularly Black Cockatoos" (Avicultural Magazine 1949 page 31).

From this information I feel we must credit Sir Edward Hallstrom with the world first breeding of the Red-tailed Black Cockatoo in 1943.

The first documented breeding of this species occurred in Adelaide Zoo in 1945 and was reported in The Royal Zoological Society of South Australia's "Zoo News" January-February and March-April 1945. Alan Lendon wrote an interesting report of the breedings of two pairs during 1945 in Adelaide Zoo in the Avicultural Magazine 1950, page 129.

An abbreviated account of this report is as follows: "The first pair consisted of a male of one of the large subspecies who was at least 35 years old and the female was of the southernmost population of the subspecies *samueli* who had been in the zoo for eight or nine years. Two eggs were laid, one proving fertile; only the female incubated and a chick was first seen on 3 February. The baby flourished and was observed looking out of the nesting log on 12 April, but it did not fledge until 10 May. This was 96 days after hatching. It was not seen to feed itself until 8 September which was four months after fledging. The youngster appeared to be a male from the start and attained full adult plumage four years later, after a series of incomplete moults.

"The second pair were younger, of the Western Australian subspecies *naso*. Their first egg was broken on the aviary floor but a second egg was laid a month later in a hollow scratched out on the ground in a corner of their aviary. This egg was incubated by the female and hatched on 5 April. The chick grew rapidly and soon began to wander around the aviary floor; this was prevented by placing bricks around the chick in the original nesting corner. The first flight to a perch was made on 18 June when the youngster was 75 days old. Like the first fledgling it was not seen to feed itself until four months after flying."

Red-tailed Black Cockatoos are the most commonly kept and bred species of Black Cockatoo in Australia. Some pairs of this species are extremely prolific. Bob Lynn has a pair of *C. m. magnificus* which have nested consistently every seven months for over 20 years and have flown in excess of 30 magnificent parent-reared young.

The dietary requirements of the Red-tailed Black appear to be more flexible in captivity than other members of the genus. This is probably due to the harsher and more variable habitat frequented by this cockatoo having promoted less specialised feeding requirements. Bob Lynn obtained his first Red-tails in 1959 and achieved his first successful breeding in 1966. The pair were housed in an aviary approximately 9 m (30 ft) long, 2 m (6 ft 6 in.) wide and 2.4 m (8 ft) high. It faces north, has a fully enclosed shelter on the southern end and is totally protected on the western side and roofed for 3 m on both ends. The nest is a vertical log similar to those used for all Bob's Black Cockatoos. The egg was laid on 4 April, 1966, and duly hatched on 4 May, 1966, after a 30-day incubation period. Feeding of the baby in the log was carried out by the female only and after 84 days in the nest the youngster fledged.

I assembled the following information on Perth Zoo's first breeding of the western Red-tailed Black Cockatoo, *C. m. naso* from extensive data compiled by Neil Hamilton and Peter Brindley. It was made available to me by the Zoo's Director, John De Jose.

The pair were housed in an aviary 9 m (29 ft) long, 1.8 m (6 ft) wide and 2.3 m (7 ft 6 in.) high, which included a 2 m (6 ft 6 in.) shelter area on the rear end. A nesting log 90 cm (3 ft) high, with an internal diameter of 45 cm (18 in.) and a natural entrance spout was provided although the female preferred to enter through the open top end of the log instead of the spout. The bottom of the log is closed with a perforated flat iron sheet and protection from the weather is provided by a sheath of flat iron fitted around the log and a sheet of flat iron above it. In the open area of the aviary, 3.5 m (11 ft 6 in.) from the back wall, there is a metal stand 1.2 m (4 ft) high on which the log stands in a vertical position. Natural wood dirt and chips are placed in the log to a depth of 15 cm (6 in.).

The first egg of 1987 was laid on 13 January and found broken nine days later. On 26 March two infertile eggs were removed from the log, which was then relocated further from the front of the aviary to the less stressful position indicated above. Soon after, on 3 April, the hen was presumed to be incubating an egg. The female's behaviour at the entrance of the log on 1 May gave the impression that the egg had been hatched. An inspection of the nest on 1 June revealed a baby about four weeks old and on 16 June the youngster was heard being fed for the first time.

Large quantities of Casuarina cones and very little green food were being consumed at this stage. The youngster was observed looking out the log entrance for the first time on 7 July. On 9 July the male was seen feeding the baby at the mouth of the log while he perched on top. He then immediately fed the female on three distinct and consecutive occasions. Next day the youngster was sitting at the top entrance to the log and the following day on top of the log. On 12 July the baby left the nest, approximately 73 days after hatching. It was first seen hanging on the wire mesh of the aviary and then later in the day was seen to fly the width of the aviary and back again. The following morning it was found on the floor in a weakened condition and replaced in the log. Both parents still appeared attentive towards it. When examined by the zoo veterinarian the same afternoon it was decided to remove the fledgling for hand rearing. A mixture of four parts wheat germ, three parts mixed rice cereal, two parts Farex, two parts Creamota, two parts ground wheat and oats to which was added vitamin drops, was fed as required daily. Every second day a mixture of Complan, finch mix, turkey starter pellets and calcium was added to the basic mix.

This youngster was eventually deemed independent on 21 December, 1987.

Three more eggs were laid during 1987, in August, September and October, but the female refused to incubate them.

Sexing

Adult Red-tailed Black Cockatoos are readily sexed, the male being basically black in colour with a broad bright red bar across the lateral tail feathers (all main tail feathers with the exception of the two central feathers). The beak is black. The female is brownish black with small yellow spots on the head, neck and wings, together with the barring ranging in colour from pale yellow to bright orange on the front, underparts and tail. The beak is horn coloured.

Immature birds are similar to adult females with a darker beak, although odd youngsters sometimes leave the nest showing slight indications of their sex. These indications can be misleading at times. Generally speaking young Red-tails are difficult to sex until three years of age, when the male's beak darkens the yellow spotting starts to disappear and some of the red barring may be visible in the tail feathers. Full adult plumage is attained at four years of age and sexual maturity is reached at the same age.

Display

The male approaches the female in the usual excited manner, bouncing along the perch with his tail fanned, his crest erected and extended fully forward. The Red-tail's crest appears to be capable of extending further forward than other members of the genus. The head feathers are puffed out to almost hide the beak, while he bows periodically. The display is usually accompanied by a muffled gurgling call and followed by the male feeding the female.

Nests

Again Bob Lynn favours vertical hollow logs standing on the ground under shelter which measure approximately 1.5 m (5 ft) long and 30 to 40 cm (12-16 in.) internal diameter with a natural entrance spout. (See Plate No. 1B.) They are filled to within 60 cm (2 ft) of the top with rotten wood, dirt or soft termite nest. Bob fills the logs to this depth to allow easy examination of the nest and has found that most Black Cockatoos will accept the depth of the nest provided, as long as it is not too deep. They will not usually excavate the hollow deeper as many white species will do.

Other breeders have had success with suspended logs of similar dimensions in both vertical and horizontal positions.

Nesting and Hatching

Breeding has been recorded throughout the year in New South Wales aviaries, but it does appear to be more seasonally orientated in the southern states of Australia. Mark Schmidt states that his Red-tails usually breed from January to April in South Australia.

The intention to breed is indicated by constant chewing activities around the nesting log for some weeks prior to the egg being laid. Usually one and very occasionally two eggs are laid and incubated by the female. She is fed by the male during incubation and for two to three weeks after hatching. Incubation periods of 28, 29 and 30 have been recorded with a predominance of 30-day periods being recorded in Bob's aviaries. Peter Hobbs of New South Wales has recorded one incubation period of 31 days and two of 32 days from his pair of *C. m. naso*. The chicks hatch with bright yellow down, which gives them the appearance of a yellow ball of fluff. When on rare occasions two chicks are hatched, the youngest is neglected and allowed to die. I can find no records of Red-tails rearing a two-young clutch.

At two to three weeks of age the babies eyes start to open and early pin feather development can be observed. Also at this stage the female ceases to brood the chick during the day, except in periods of inclement weather. All feeding in the nest is carried out by the female.

tling three days old.

B Nestling three weeks old.

tling five weeks old.

D Nestling six-and-a-half weeks old.

E Nestling 11 weeks old.

Plate 13. Red-tailed Black Cockatoo — Nestlings.

Fledging periods of 75 to 96 days have been recorded. This large variation can be related to climatic conditions, diet, the feeding ability of the parents and the precociousness of the individual young. Both parents feed the youngster for up to three to four months after leaving the nest. The male often carries out the majority of this duty. When the parents cease to feed the young after three to four months it is normally safe to remove it from the aviary. Young Red-tails leave the nest similar in appearance to the adult female but with a light grey beak and noticeably paler grey eye rings which darken similar to the adults by twelve months of age. When young are being parent reared Bob feeds his normal diet but increases the ration of Casuarina cones considerably.

We have recorded fresh egg weights of 29.95 g, 30.1 g, 30.84 g, 31.15 g, 32.42 g, 33.32 g, 34.9 g and 37.9 g for *C. m. naso* as well as 37.74 g and 38.5 g for *C. m. samueli.*

Mutations and Hybrids

There are no records of mutations occurring in this species. The only hybrid records are between the subspecies *C. m. magnificus* and *C. m. macrorhynchus.* (Prestwitch — Records of Parrots Bred In Captivity) and *C. m. magnificus* and *C. m. Samueli* (Allan Lendon: "Australian Parrots in the Field and Aviary" 1973).

GLOSSY BLACK COCKATOO

Plate 14. Glossy Black Cockatoo *Calyptorhynchus lathami* — Male.

GLOSSY BLACK COCKATOO

(Calyptorhynchus lathami)

Derivation:

Calyptorhynchus:

calyptos — Greek for hidden

rhynchus — Greek for beak

lathami — after the English ornithologist John Latham

Description:

Length 50 cm

Weight 425 g average

Appearance as in Plates Nos. 14 and 15

Classification

The unique mandibles of this species which are perfectly designed to hold and extract the seeds from the Casuarina cone, reflect its very specialised feeding habits, and set it aside from all other members of the genus. Both mandibles are particularly broad, the upper protrudes in a balloon like fashion terminating in a small sharp point, while the lower mandible has two well separated, tooth-like protrusions which grip the cone while the upper mandible tears it open to expose the minute seeds. All the birds of this species which I have been able to observe closely when feeding, held the cone in the right claw while removing the seed from the cone. When the seed has been freed from the cone the lower mandible is moved to the bird's left thus bringing the right side protrusion in line with the point of the upper mandible enabling the seed to be split and the kernel extracted.

There are no subspecies recorded in this distinctive, highly specialised and smallest member of Calyptorhynchus, although data collected by the RAOU for their "The Atlas of Australian Birds" suggests three distinct populations. These consist of the main population extending along and eastward of the Great Dividing Range, a western population and a small isolated population on Kangaroo Island. The total similarity of the isolated Kangaroo Island group to the other populations and the range of this population suggests to me that the isolation has occurred in comparatively recent times and that this cockatoo originally ranged along the Darling and Murray River systems, but changed conditions, possibly in food source,

Plate 15: Glossy Black Cockatoo *Calyptorhynchus lathami* — Female.

forced the retreat of this species in both directions and thus isolated the south-western section of the population.

Earliest Report

Unfortunately the early references to this species are confused as it was considered by some ornithologists to be an immature Red-tailed Black Cockatoo and by others to be a small subspecies of the Red-tailed. This species was illustrated and described in Phillip's "Voyage to Botany Bay" page 267 with the plate opposite page 152 in 1789, but incorrectly named the Banksian Cockatoo. White also illustrated and described this species and also incorrectly named it the Banksian Cockatoo in his "Journal of a Voyage to New South Wales" 1790 on plate opposite page 139.

The confusion was eventually solved and the species named *Psittacus lathami* by Temminck in "Catalogue Systematique du Cabinet d'Ornithologue" in 1807, page 21, after the English ornithologist Latham, who had named and described the Red-tailed Black Cockatoo, but whose name *Psittacus banksii* was eventually not accepted.

Range, Habitat and Field Notes

The Glossy Black Cockatoo ranges over eastern Australia, from the mountain ranges west of Mackay, Queensland, in the north to the Bairnsdale area in Victoria in the south. Then west to the Bolon-St George region of Queensland and the Griffith area of New South Wales. As well there is the isolated population on Kangaroo Island. Birds from this isolated population make irregular appearances on the adjacent Australian mainland south of Adelaide. In the northern parts of the range this species is usually confined to elevated forest regions whereas elsewhere it is widespread.

All my field observations of this Cockatoo have been in or close to Casuarina trees and usually of pairs or small parties feeding on the seeds of Casuarina cones.

The sighting that remains most vivid in my mind is that of a pair feeding in a Casuarina tree on the edge of an extensive picnic area at Bobbin Head in Ku-ring-gai Chase National Park which is situated on the northern fringe of suburban Sydney. I was attending a family picnic one Sunday in 1981 and the picnic areas of the park were alive with hundreds of people enjoying the recreational activities associated with picnics in Australia. The noise associated with the people was, of course, quite deafening at times. While we were relaxing on the grass, enjoying a mid-morning cup of tea, I noticed chewed Casuarina cones falling at regular intervals from a large Casuarina tree which stood about 10 metres high and grew from near the base of a steep rocky hill. A closer inspection revealed two Glossy Black Cockatoos feeding casually near the top of the tree. They fed silently and were quite indifferent to the constant movement on the ground below. I climbed the almost cliff-like hill adjacent to the tree until on a level with the Cockatoos, thus providing myself with a clear view of the birds, at a distance not exceeding 12 metres. The Glossys, which I could now see were an adult pair, were undisturbed by my presence and continued their routine of snipping off a cone, holding it in their claw while extracting the tiny seeds with their beak and dexterous tongue, then discarding the cone and taking another. I watched these birds for well over an hour before being called for lunch and then kept my eye on them for a couple more hours from the picnic area below and not once did I hear them utter a sound. Later in the afternoon they flew silently away, unnoticed by every person in the park — except one.

This Cockatoo is not always a quiet feeder, perhaps the obvious human presence motivated them to remain silent.

Sightings of this species in rural suburbs on the outskirts of Sydney are still not uncommon although their numbers have obviously declined considerably in the Sydney region over the last 50 years. Glossys are probably more abundant in south-eastern Queensland and northern New South Wales than in other parts of their range. In some localities they could be termed almost common, yet in other areas of apparently suitable habitat they are scarce. Some regions have resident populations, whereas in others they are seasonal visitors and in yet other localities they are nomadic wanderers, occurring only infrequently, all of which must be related to the availability of suitable food.

The habitat frequented by this Cockatoo is coastal and mountain Eucalypt forest country, dry open forest and trees adjacent to lakes, creeks and rivers, and it is always dependent on the presence of Casuarina trees.

Their diet consists almost entirely of Casuarina seeds which is occasionally supplemented with Angophora, Eucalypt and Acacia seeds, wood boring insect larvae as well as some seeds from other native trees and shrubs. This species is commencing to sample the seed from introduced flora and has been observed feeding on the seeds of the Japanese White Cedar tree and ripening sunflower seeds. Both observations occurred in rural areas adjacent to Sydney. The Glossy Black Cockatoo is exclusively an arboreal feeder and comes to ground only to drink.

The flight of this species is unlike that of any other member of the genus and is best described as slow floating flight with short wing beats.

Breeding in the Wild

Breeding usually takes place during the autumn and winter months. The preferred nest sites are a large hollow limb, a hollow in the tree trunk where a branch has broken off, or a tall stump of a snapped off tree, usually in Eucalypts or sometimes a dead tree at heights of 9 m (30 ft) to 20 m (65 ft). The nesting site is prepared by the male and female by chewing at the inside walls of the hollow and dropping the splinters and chips to form the nest base. There are odd reports of nests being lined with Eucalypt leaves, similar to that of the Galah. These are supported by at least one case of nest lining in an aviary situation, although this practice is obviously the exception, not the rule.

A single white oval egg is laid and incubation is usually commenced immediately by the female who sits alone. She is called from the nest during incubation and for the first week after hatching to be fed by the male in a nearby tree. After being fed the hen returns to the hollow and enters tail first. As the chick develops the female spends less time in the nest and broods the chick only at night and during periods of cold weather. When the chick has grown sufficient feathers all brooding ceases and the hen often roosts on or near the entrance to the nest at night. At this stage the female spends most of her time feeding with the male, usually returning to the nest in the morning and late afternoon to feed the chick. Fledging occurs about three months after hatching, then the youngster is fed by the parents for several months.

After breeding, family parties and small groups are usually formed, but on occasions large flocks of 40 or more birds have been recorded.

The following field information relating to the nesting habits of Glossy Black Cockatoos in a particular locality of northern New South Wales were given to me by a friend of the observer whose expertise and local knowledge of this species is remarkable — many thanks.

In this region the Glossy Black Cockatoo usually selects a hollow in an isolated Blackbutt tree (of nine recorded nests, eight were in Blackbutts). All nest site trees stood in small clearings, surrounded by low Casuarina forest and were close to fresh water. The nesting hollows were all in dead vertical spouts, averaging 22.5 cm (9 in.) internal diameter and standing clear of all foliage and thus receiving full sun during the entire day. Heights of the nesting hollows ranged between 18 and 28 m (60 and

90 ft) above the ground. Entrance to the nest chambers were straight down the vertical hollows. The depths of the nests varied from 60 to 90 cm (2 to 3 ft) and all nests examined had a base of chewed up wood.

One egg was laid during March or April and while the females incubated the males were absent from the nest site areas. The females fed for short periods on Casuarina seeds taken in close proximity to the nest site during incubation and for several days after hatching as well as being fed by the male. Ten to 14 days after the chick hatches and during sunny weather the females leave the nest just after sunrise and do not return to feed and brood until almost dark. Just prior to entering the nest hollow the female squawks very loudly.

In this region the Glossy Black Cockatoo breeds in loose colonies with three nests having been recorded within a half-mile radius.

The observer wishes to remain annonymous to protect the nest sites.

Aviculture

Early avicultural information is non-existent for this species, apparently due to the difficulties encountered by early aviculturists to induce these birds to eat a domesticated diet. Sir Edward Hallstrom appears to have been the first aviculturist dedicated enough to pursue the dietary problem to a satisfactory conclusion. His early experiments were carried out on adult birds who were fed a limited supply of natural diet (Casuarina cones) which Sir Edward considered necessary not only for their nutritional value, but for providing the chewing required to maintain the beak in good condition. These were supplemented with a mixture of Canary seed flour, peanut butter and peanut oil administered direct to the crop with a syringe and rubber tube. When Casuarina cones were not fed both the top and bottom beaks soon became overgrown. The ultimate aim was to induce the Glossys to live successfully on a diet of limited Casuarina cones and sunflower seeds. This was accomplished after six years of perseverance. Sir Edward tried unsuccessfully to introduce additional food items, however the birds survived and eventually bred on this diet.

Sir Edward Hallstrom achieved the world first breeding of the Glossy Black Cockatoo in 1954. The following is an abreviation of his account which was published in the English Avicultural Magazine Volume No. 60 Page 163.

Two pairs were set up for breeding in separate large aviaries and provided with suitable hollow logs. One pair was seen mating yet the second pair, who were friendly enough but did not seem to be interested in breeding, soon laid one egg. Apparently this egg was not incubated for the account goes on: Several days later a second egg was laid during April 1954 and duly hatched after a 29-day incubation period.

Soon after it was noticed that the male was loosing condition, which Sir Edward reasoned was due to him starving himself in his efforts to feed the female, who in turn fed the young and was seldom seen out of the log. The male was removed from the aviary and force-fed direct to the crop with the canary seed flour, peanut butter and peanut oil mixture. After a few days he showed good progress and was returned intermittently to his aviary with the hen and baby. But the double feeding responsibility proved too much for him, even though he had an unlimited supply of Casuarina cones, and he had to be removed again.

At this stage — 1 June, 1954 — the baby was over two weeks old and was the size of a clenched fist and covered with a fluffy cream down. The baby continued to do well and by 25 June, 1954, was almost fully feathered and showing itself at the entrance of the nesting log. As the female was still carrying out all rearing duties quite successfully on her own, Sir Edward did not return the male for fear of disturbing her.

On 20 August, 1954, the young Glossy left the nest with a considerable amount of long down protruding between the feathers. It was soon accompanying the mother to the food dishes and partially feeding itself and was eventually reared to total independence solely by the female.

"Some Further Breeding Results in the Hallstrom Collection" published in the English Avicultural Magazine 1959, states that in 1958 another Glossy had been successfully reared and that two previously reared youngsters were now mature and showed an interest in breeding. It was also stated that the original pair was incubating on two eggs, which was the first time a two-egg clutch had been seen in the Hallstrom aviaries.

Lendon states that at the time of Sir Edward Hallstrom's death in 1968, there were 13 adult Glossy Black Cockatoos in the collection and that Gus Beilharg, the keeper, had told him that several young had been successfully reared over the years, but all from the one pair. Other pairs had often laid, but the eggs were always infertile.

Taronga Park Zoo succcssfully reared a chick in 1970 from an exhibited pair, which originated from the Hallstrom Collection.

Kerry Muller, a past Curator of Birds at Taronga Park Zoo, stated in "Australian Birdlife" Volume 1, No. 6, pages 116 and 117, that the Glossys in Taronga nested in large hollow logs which were open at the top and hung at a 45 degree angle about one metre from the ground. A single egg clutch is laid, usually in August and incubated by the female for 28½ to 29 days. The newly hatched chick is densely covered with sulphur coloured down and has blackish unopened eyes and black beak. Kerry also records a fledging period of approximately 60 days, which is not consistent with other records of the time spent in the nest by Glossy chicks.

Bob Lynn obtained his first Glossy Black Cockatoos from Sir Edward Hallstrom, and they laid their first egg in his aviaries on 17 March, 1968, which eventually proved to be infertile as were several other eggs laid during the following three years. The first fertile egg was laid on 31 March, 1971, and was hatched on 30 April, 1971, after a 30-day incubation period. The chick was covered with yellow down and was fed only by the female, who in turn was fed by the male, usually at the entrance to the log. Development of the chick was rapid; at three weeks old the eyes were open and early pin feathers were apparent and by eight weeks old the chick was well feathered. Fledging occurred on 22 July, 1971, 83 days after hatching and the chick was fed by both the male and female. The male, however, did not show the interest in the youngster normally shown by males of other members of the genus. Bob has found over the ensuing years that all his cock Glossys have shown only a waning interest in their young after fledging and with some males this soon turns to aggression.

During 1985 I took a series of nestling photographs of a Glossy Black Cockatoo chick in Bob's aviaries, some of which appear within these pages. The chick developed into a beautiful, robust youngster and when almost fully feathered was observed to have large blotches of yellow feathering on the head, which is indicative of a hen, but usually only appears as maturity approaches and is most unusual to be seen in a fledgling. When the chick fledged even more yellow was apparent, which convinced Bob that the bird must be a hen.

Another interesting aspect of this particular breeding was that the hen showed a marked preference for fresh green peas as opposed to the normally preferred Casuarina cones, when rearing young. The chick was almost totally reared on green peas, which must represent a major break-through for the adaptation to a totally domesticated diet for the Glossy Black Cockatoo. Sprouted corn, sprouted sunflower seed, almonds and raw peanuts were also available during the rearing period.

After a few months this youngster had to be removed from the parents due to the aggression shown towards her by the father. She was placed in an aviary with a four and a half year old male who was delighted with her company; they settled down quickly and appeared quite compatible.

Two seasons later in 1987, when the young hen, who now displayed more yellow on the head than an average adult female, was just over 18 months old and the cock six years old, the pair showed interest in a nesting log. This log stood vertically in the roofed area of the flight and was 1.4 m (4 ft 6 in.) high and 30 cm (12 in.) in diameter. Everything about this young hen is unusual, for she immediately commenced to line the nesting chamber with Casuarina needles (leaves) and by the time the egg was laid a definite nest had been formed. (See Plate No. 16B.)

The egg was laid on 29 March, 1987, when the hen was 20 months old; incubation commenced immediately and the chick hatched on 28 April, 1987, after a 30-day incubation period. I photographed the chick on 7 May, 1987, when it was nine days old, but unfortunately two days later a freak driving storm flooded the log and the chick was drowned, even though it was under shelter.

Sexing

The Glossy Black Cockatoo differs from the other Red-tailed members of the genus in that the male and female have similar basic body colouring and the male's deep grey beak is usually only marginally darker than the female. The male has a broad bright red bar across the lateral tail feathers, whereas the female's tail bar ranges in colour from pale red to orange then to yellow and is crossed by narrow black stripes while the head and neck is mottled with varying yellow spots and blotches. Very occassionally hens have been seen with full yellow heads, while conversely nesting females in the wild have been observed without any yellow on the head.

Immatures are similar to adult females without the yellow head and neck blotches, but with varying amounts of yellow spotting, similar to the female Red-tailed Black Cockatoo, on the upper and under-wing coverts, head, ear coverts and at times on the side of the body under the wing. There is some variable yellow barring on the throat and body; the lateral tail feathers have a broad red and yellow band crossed by five or six narrow black bars.

A guide to the sex of birds under 18 months old can be gained by observing the colour of the spotting on the underwing coverts and on the side of the body under the wing, which are usually orange red in males and yellow in females. Usually the spotting and barring has moulted out by 18 months of age, although some retain them until the adult moult. Young males' red and yellow barred tail feathers are sometimes replaced with one or more red unbarred feathers on the second moult.

The immature sex differences are only guides to sex and must not be regarded as infallible. It is generally accepted that Glossys can be sexed at three years of age, although Bob Lynn has had some birds whose sex was uncertain until three and a half years old.

Display

Mated pairs of this species indulge in mutual preening and males will often display to the female throughout the year, but as the breeding season approaches, display together with courtship feeding becomes increasingly more regular.

The display consists of the male approaching the female with his short crest erected, his entire head feathers puffed out at right angles (in a manner I find reminiscent of the display of the South American Hawk-headed Parrot, *Deroptyus accipitrinus*, on a much reduced scale) and tail fanned to show the red bar, while bowing, bobbing his head and emitting a soft low whistling call.

Copulation was described by Neville and Enid Connors in Australian Aviculture May 1988 as follows; Mating usually occurs in the morning with the hen positioning herself on the perch with tail raised and wings dropped. The male mates with both feet on her back and with his wings fully open. He then dismounts and flys off to the end of the aviary before returning to repeat the performance three or four times.

Nests

The most successful nesting site appears to be a vertical hollow log approximately 1.4 m (4 ft 6 in.) high, with a natural looking entrance and an internal diameter of approximately 35 cm (14 in.) and standing on the ground. Similar sized logs hung vertically or on a 45 degree angle with the bottom sealed, have also been used with some success. Neville and Enid Connors provide nesting logs approximately 1 m (39 in.) long with an internal diameter of about 30 cm (12 in.) and hung in a vertical or near vertical position with an access door near the base. The most favoured nesting materials are rotting wood dirt, crushed soft termite nest or wood shavings, which should fill the hollow to within about 60 cm (2 ft) from the entrance. This is a convenient depth for a nest, giving accessibility within arm's length.

Nesting and Hatching

Prior to nesting the male and female chew around the nesting log, dropping chips and splinters of wood into the nest and of course increased display and feeding of the hen by the male is observed during this period. Bob Lynn has recorded eggs laid between 1968 and 1988 in February (one season only) March, April, May, June and on one occassion in September. Kerry Muller stated that the Glossys at Taronga Park Zoo usually laid in August. John Courtney recorded eggs being laid in March and May (The Australian Bird Watcher, Volume 11, No. 8, 1986) and Sir Edward Hallstrom recorded an egg being laid in April. Neville and Enid Connors of northern New South Wales record breeding as early as February and as late as October with April and May the usual months when egg laying occurs (Australian Aviculture, May 1988).

The vast majority of eggs laid by Glossys in New South Wales aviaries are during the autumn months. One egg (very rarely two) is laid and incubated by the female, who is fed by the male during the incubation period and for the first few weeks after hatching. The occasional recorded case of two eggs being laid by this species appears to be due to an aborted nesting on the first egg with the second egg being laid several days later. In many instances hen Glossys abort their first egg of the season then lay a second egg a few weeks later which they incubate. Hallstrom, Lynn and Connors all record this behaviour.

We have recorded fresh egg weights of 16.25 g, 18.95 g, 22.43 g and 22.53 g.

Incubation periods of 28½ days (Muller and Connors), 29 days (Hallstrom, Muller, Lynn and Connors), 30 days (the usual period recorded by Bob Lynn) and 33 days approximately (Courtney) have been recorded. Jim Gill also reports an artifical incubation period of 32 days. I consider all these incubation periods to be genuine records. The shorter periods from Hallstrom, Muller and Connors were recorded close to the coast in the Sydney area and on the north coast of New South Wales, which must be considered a mild climate with a low day-night temperature variation, whereas Bob Lynns records are taken 25 kilometres (15 miles) further west (inland) in the Sydney region where the temperature variations are more extreme. John Courtney's records were taken in the much colder New England area of New South Wales where the excessive temperature variations may well promote longer incubation periods.

The chick hatches with thick long yellow down and horn coloured beak. Usually by three weeks old the eyes are starting to open and the early pin feathers are appearing beneath the skin. At 10 weeks the young is almost fully feathered and fledging usually occurs between 75 and 90 days. Although Muller recorded a fledging period of 60 days at Taronga Park and Courtney recorded a fledging period of about 100 days in the

New England region, both records must be considered extreme cases. Neville and Enid Connors record an average fledging period of 90 days.

The young are fed by the parents for three to four months after leaving the nest, but Bob Lynn states that his male Glossys assume less of this responsibility than males of the other members of the genus and that some males become quite aggressive towards their young.

Youngsters leave the nest as previously described with a considerable amount of the hair-like down still visible and with a paler grey beak and eye ring.

Fledglings can normally be removed from their parents four months after leaving the nest, but if the male is excessively aggressive it may be necessary to remove the youngsters sooner. Should this be the case supplementary feeding may be necessary until the baby is fully independent.

Sir Edward Hallstrom stated in his accounts of the dietary problems of the Glossy Black Cockatoo published in the Avicultural Magazine (England) that this species does not have a gizzard similar to the other Black Cockatoos. I have been unable to find evidence to justify this statement as the anatomy of this species is similar to the other Calyptorhynchus. Probably the variation noted in the gizzard of the Hallstrom Glossy was related to the cause of death of the birds autopsied.

John Courtney in "The Australian Bird Watcher" Volume 11, No. 8, page 272, 1986, denounces the absolute necessity to feed Casuarina seed to the Glossy Black Cockatoo. His Glossys and those of many other aviculturists of his acquaintance have never been fed Casuarina seeds. John's birds, who, by the way, have successfully reared young, are fed sunflower seed, milk arrowroot biscuits, Nutri Grain breakfast cereal and assorted nuts including red skinned raw peanuts.

It appears that the pioneering days in avicultural techniques for this species may almost be over, although the more successful breeders of this species, such as Bob Lynn and Peter Chapman still supplement their diet with Casuarina cones. This natural food is becoming increasingly more difficult to obtain around Sydney, due to increased National Parks and Reserve Areas, contamination from vehicle exhaust fumes, the regular burning of scrub land by Bush Fire Brigades and last but by no means least the continuing spread of suburbia.

Many innovative aviculturists are experimenting with totally domesticated diets with some success.

The following daily growth rate chart compiled by Neville and Enid Connors and published in Australian Aviculture May 1988 provides invaluable information which may be used as a comparison when hand rearing this species.

Day	*Weight* *g*	*Notes*
1	15.12	
2	14.58	
3	15.25	
4	16.28	
5	18.79	
6	18.79	
7	21.45	
8	21.80	
9	25.63	
10	30.61	
12	37.78	
14	47.81	Droppings more solid, eyes slit partly open
16	56.17	
18	67.02	Pin feathers through skin particularly on chest
20	75.43	Eyes wide open, pin feathers all over
22	87.63	
24	96.78	
26	106.74	Pin feathers on chest breaking open
28	112.27	
30	130.80	Closed ring fitted
32	137.10	Feathers on chest
34	144.60	Feathers around beak
36	144.20	Tail feathers appearing
38	156.50	
40	162.20	Showing some aggression
42	167.90	Many black feathers, still much down
44	188.90	
47	177.30	Mostly feathers showing, still showing down
50	184.80	
54	197.10	
58	205.20	Yellow orange spots under wings
62	221.70	
65	230.60	
70	246.80	
76	258.40	
94	290.20	Flying

The Connors went on to say that this bird proved to be a hen and that a cock they reared weighed 325.90 g at fledging, while another youngster of unknown sex and who had been ill during the rearing period weighed 265.80 g when it flew.

Mutations and Hybrids

I can find no records of a colour mutation occurring in this species. However, I recently received information relating to a young Glossy Black Cockatoo held in a New South Wales aviary, which has an orange band edging every feather. Whether this colour variation will be retained after the adult moult is problematical.

There is no record of hybridisation occurring with this species.

tling four days old.

B Nestling nine days old in hollow lined with Casuarina needles.

tling 10 days old with hen in foreground.

D Nestling six weeks old.

E Nestling 11 weeks old just prior to fledging.

Plate 16. Glossy Black Cockatoo — Nestlings.

GENUS CALLOCEPHALON

Plate 17: Gang Gang Cockatoo *Callocephalon fimbriatum* — Male and Female.

GANG GANG COCKATOO

(Callocephalon fimbriatum)

Derivation:

Callocephalon:

callos — Greek for beauty
cephale — Greek for head

fimbriatum:

fimbriatus — Latin for fringed

Description:

Length 35 cm
Weight 300 g
Appearance as in Plate No. 17

Classification

The Gang Gang Cockatoo is the only member of this genus (i.e., it is a monotypic genus) and differs noticeably from all other Australian Cockatoos with its unusual flimsy, fine feathered crest, which curves forward. It could be described as an overflowing crest, particularly in a fully matured male. The powerful, heavy upper mandible is deeply notched to accommodate the tooth-like protrusions on the lower mandible, which are an adaptation for opening hard seed pods. The legs are short and thick and the feet are very dexterous as is usual in psittacines with an arboreal lifestyle.

One subspecies, *Callocephalon fimbriatum superior*, from the Blue Mountains in New South Wales, was recorded by Matthews and was said to be of paler colouration. It is now considered that this subspecies is not valid and related to a few palid individuals only. The limited range and the constant intergration due to this cockatoo's nomadic habits has prevented the formation of authentic subspecies.

Earliest Reports

This cockatoo was described by Latham in his Second Supplement to his "General Synopsis of Birds" in 1801 and was named the Red-crowned Parrot, *Psittacus galeatus*. The first published illustration of this bird was in Grant's, "The Narrative of a Voyage of Discovery" 1803 (plate opposite page 135) under the scientific name of *Psittacus fimbriatus*.

Range, Habitat and Field Notes

The Gang Gang Cockatoo ranges approximately from the Dorrigo region of New South Wales through the south-eastern coastal, mountain and highland country of New South Wales, the whole of eastern and southern Victoria and the south-astern corner of South Australia. Surveys carried out by the RAOU for "The Atlas of Australian Birds" indicate that the Gang Gang is now absent from Tasmania and King Island and is scarce on Kangaroo Island where it was introduced.

This cockatoo is mainly a bird of forest country and is often observed in coastal and mountain forest regions where it follows a nomadic lifestyle which is solely regulated by food supply. They are commonly seen in more open rural areas of the Southern Tablelands of New South Wales during the autumn months, feeding on the ripening berries of the introduced Hawthorn Bush *(Photinia serrulata)*. Gang Gangs are often remarkably quiet when feeding and will allow close observation, particularly during the autumn and early winter, when many recently fledged young are in the flocks. This is a vulnerable period for these immature birds with many falling prey to predators, both human and otherwise. The practice of catching young Gang Gangs during this period by pulling them from a feed tree with a snare on the end of a long rod is common among bird trappers.

All bird orientated people, regardless of the area of their interests, must combine to protect this and other vulnerable species during these susceptible periods. I have found that later in the year, as the youngsters become more experienced and their search for food takes them into more remote regions, they are less vulnerable.

Over 30 years ago I witnessed a disgraceful scene when a bird was shot from a flock of fifty or more Gang Gangs which were moving noisily through some low forest country south-west of Sydney. The bird was only wounded and fell squawking to the ground and the flock immediately congregated around it, swooping low over it and some even landing on the ground beside it, all calling and showing tremendous concern for the fallen bird. The flock did not move away until the wounded bird was killed and I am sure that if the shooter had so desired he could have shot the entire flock, like ducks in a shooting gallery.

This again demonstrates the need for concern for the welfare of this adorable species, whose very disposition jeopardises its continued existence.

On a brighter note I would like to relate to a pleasant afternoon spent in the National Botanical Gardens of Canberra, A.C.T., which contains only native flora, plus a resident population of Gang Gangs. Joe Forshaw had taken me there to see the Gangs, knowing they are my favourite bird and that watching them in the wild never fails to give me much pleasure. After wandering these beautiful gardens for an hour or so, we were rewarded by the distant calling of Gang Gangs as they moved through the tree tops towards us. Soon three or four pairs were close by and an immaculate adult pair landed in a large Eucalypt just above us. Although it was mid-winter the male flew directly to a hollow limb and started to examine it, while calling excitedly to his mate who flew closer and watched. The male remained on top of the entrance spout, chewing at the bark and clucking to himself in that distinctive Gang Gang way, for quite some time.

I have seen Gang Gangs in the field on numerous occasions but I am still just as thrilled and enthralled by a sighting of this species as I was the first time I saw it.

My friend and veterinarian, Jim Gill, maintains that his most pleasureable bird watching experience was with Gang Gangs. He was on horseback mustering cattle in the Rylstone district of New South Wales. As he was riding through a lightly timbered Eucalypt forest a mature pair of Gang Gangs flew alongside him for what seemed like ages leaving him with an everlasting memory of their beauty and a sense of unity with them in nature.

The presence of this cockatoo in an area is usually first determined by the sound of its unusual grating, clanging call which cannot be mistaken for any other bird in the Australian bush. Their flight is slow and laboured and has often been described as owl like.

Their diet consists of seeds, fruit, nuts and berries of native trees and shrubs, such as Eucalypt and Acacia and of introduced flora such as Hawthorn *(Crataegus monogyra)*, Firethorn *(Pyracantha species)* Cotoneaster (Cotoneaster Species) and various pines. Vegetable matter, insects and insect larvae are also consumed.

The following personal observations related to me by John McGrath of Yass, New South Wales, are of great interest.

John has been observing the flocks of Gang Gangs which arrive annually in his district during late summer or early autumn for many seasons.

Apparently soon after the young have fledged, flocks are formed and congregate in this locality at a time coinciding with the ripening of the native Stringy-bark seeds and to a greater extent the berries of the introduced Hawthorn, which form the major food source for many weeks to come.

Upon arrival, the season's young are held in a creche situation in tall Eucalypts close to the fruiting Hawthorn bushes. The creche is maintained for the first week or two until they become familiar with the locality and the food source.

This is the only report of Gang Gangs forming creches similar to Galahs that I am aware of.

Breeding in the Wild

The nesting site is usually a hole in a tree or a hollow limb, invariably at a considerable height above the ground, usually in excess of 20 metres and often in a Eucalypt tree in inaccessible places. Nesting trees have often been observed in cool secluded mountain valleys and in coastal forest areas of southern New South Wales.

As the breeding season approaches, which usually extends from October to January, mated pairs select their nesting hollows. Both male and female prepare the nest by forming a base of chewed up decayed wood in which two (very seldom three) white rounded oval eggs are laid. Incubation is carried out by both male and female, who share the task during the day, with the female incubating at night. The young are fed by both parents while in the nest and this procedure continues for many weeks after fledging. I can find no record of Gang Gangs ever being double brooded in the wild.

When breeding is completed, flocks are formed from small family parties, with up to 50 birds in one flock. These flocks, although nomadic often follow set feeding paths, moving from one regular food source to another and then another as the seasons and hence the food sources change. The annual appearance of this cockatoo in a particular area at a particular time will coincide with the ripening of a favoured food and is usually very consistent.

Aviculture

Early avicultural records for this species are scarce, but Gang Gangs were kept by English and European aviculturists during the 1880's.

The first successful breeding record was from Mme Lecallier's aviaries in France during 1921 and recorded in the Avicultural Magazine 1921, page 176 and 1922, page 7. I think an abbreviated report of this concise breeding record worth repeating.

The breeding pair comprised of an acclimatised hen from the Marquis of Tavistock's aviaries in England and the cock was a mature newly imported bird from Australia. They were placed in a newly constructed aviary approximately 5 m (16 ft) long, 4.3 m (14 ft) wide and with a height of 3.6 m (12 ft) in the centre of the aviary. This high centre section was sheltered by a 90 cm (3 ft) square roofed area under which was fitted a five-gallon cask which was covered with zinc sheeting for protection and with a 9 cm (3.5 in.) diameter entrance hole in the end of the cask.

They were released into the aviary in May 1920 and in May 1921 were seen mating and working the nesting barrel. The first egg was laid on 4 June. Incubation was shared by the hen and cock, with the male always relieving the hen for an hour or two in the afternoon. The first egg hatched on 4 July but the chick died the next day; however the second was seen chipping on 5 July and hatched on 6 July. The newly hatched chicks were covered with a rusty coloured down. At one month old the baby had well advanced blue grey pin feathers and at six weeks old red feathers in the crest denoted the baby to be a cock. When 11 weeks old he was seen to be looking out of the nest and he fledged on 4 September. A full red crest was attained at 12 months of age.

Although, as this breeding report indicated, the Duke of Bedford (Marquis of Tavistock) had Gang Gangs in his aviaries prior to 1920, apparently he did not successfully breed them until 1951, which demonstrates the difficulties which can be encountered in getting this cockatoo to breed. The first recorded Australian breeding of this species is attributed to Miss S. Merrifield of Jamestown, South Australia, in 1946.

Sexing

Gang Gangs are particularly sexually dimorphic with adult hens having orange barring on the chest and adult males having a bright red head and crest. The obvious dimorphism of this cockatoo is more advanced than in the Black Cockatoos with the sexual differences also occurring in immature birds. Although immatures leave the nest with plumage similar to but duller than the adult female, all young males have some red feathering on the head. The amount of red can be variable.

Display

I have usually observed the courtship display of this species, both in aviaries and in the wild, close to or on the entrance of the nesting hollow. The male will call excitedly to the female, often with his wings spread and his body stretched to its maximum height and his general appearance becomes particularly animated. The female will then usually fly to the mouth of the nesting hollow and examine it closely and perhaps chew a little around the entrance.

Nests

I have found that Gang Gangs usually prefer a vertical or near vertical nesting log about 60 cm (2 ft) to 75 cm (2 ft 6 in.) long and 25 cm (10 in.) to 30 cm (12 in.) in diameter. The entrance hole is preferred on the side near the top of the log, or preferably a log with a natural entrance hole or spout. When hollow logs are unavailable a nest box of similar suggested measurements and constructed of hard wood would be suitable and if chewing persists it may be necessary to protect the box with metal sheeting.

Ron Grosse is a successful aviculturist from New South Wales who bred Gang Gangs for many years and found that his birds preferred the nesting log to be inclined at an angle of 45 degrees. This has not been my experience but I would suggest that when furnishing an aviary for this cockatoo, both a vertical and an inclined log be provided, to allow the birds to make their own choice. Fine wood shavings are easily obtainable and make good nesting material. Rotten wood or peatmoss are preferred by some aviculturists.

Nesting and Hatching

In my aviaries Gang Gangs have invariably laid two eggs in the clutch with intervals of two to four days between each egg, except on one occasion when a one-egg clutch was recorded.

I have recorded egg weights of 14.25 g and 15.08 g.

Eggs have been laid in my aviaries from mid-October to mid-December. Incubation usually starts with the second egg and I have recorded incubation periods of 28 days normally and on some occasions up to 30 days. The incubation is shared by the male and female during the daylight hours and is carried out by the female at night. Chicks in my aviaries hatch with mid yellow to dark yellow down and not the rusty red down often referred to in the early European and English literature. Pin feathers start to appear and the eyes open between two and three weeks and by six weeks of age the feathering is well advanced. Fledging usually occurs about eight weeks after hatching, although this is variable as some young take a week or two longer to leave the nest. Young birds are usually fed by their parents on gradually decreasing occasions for several weeks after fledging.

Most of the juvenile plumage of the young males is replaced with adult plumage by two years of age, but complete adult plumage is not attained until at least three years of age.

I have not had Gang Gangs make any attempt to breed until they are three years old.

My first breeding of the Gang Gang occurred in 1970 with a pair housed in an aviary 7.2 m (24 ft) long, 1.2 m (4 ft) wide and 2.4 m (8 ft) high. The nesting sight was a vertical log as suggested previously. Both the male and female were seven years old. In due course a pair of young fledged in perfect condition, but my triumph was short lived, for two days after leaving the nest both babies were mutilated, through the wire partition of the adjoining aviary, which housed a breeding pair of Major Mitchells. One was already dead when I found them and the other so seriously injured that it had to be destroyed. All of my cockatoo aviaries now have solid partitions or have double wire partitions between them.

The following information has been provided by Matthew Crane of Sydney, New South Wales, and was recorded during the 1986 and 1987 breeding seasons. Matthew apologised to me for the gaps in the data provided, but no apologies are necessary. We all know the difficulties

1986 SEASON RECORDS

Day	*Weight* g	*Notes*
1 – New Years Day Chick hatched	14	
8	53	Empty crop
11		Pin feathers on wings visible
12	88	Crop three-quarters full
15	114	Crop half full, eyes open
20	164	
22	173	
25		Closed rung. Red crest feathers appearing denoting a young male
27	225	
33	265	
48	270	
49	262	
50	250, a.m. 258, p.m.	
51	246, a.m.	Supplementary feeding with 10 ml of hand rearing formula
52	241	Seven ml supplementary feeding
53	239	Ten ml supplementary feeding
54	231	Taken for hand rearing
55	230	Weight 230
56	233	
57	227	
58		Bird maintained its weight and is recovering
66		Flying strongly

encountered when holding full time employment, coping with our obligatory social activities and yet trying to give our all to such a demanding hobby as aviculture during the breeding season. Any information that time can be found to record is worth a fortune to those following us, particularly on Gang Gangs.

Matthew's pair of Gang Gangs are housed in a suspended aviary 3 m (10 ft) long, 1.2 m (4 ft) wide and 1.2 m (4 ft) high, within a sheltered area. The nesting log is 90 cm (3 ft) long with an internal diameter of 30 cm (12 in.) and it hangs externally in a vertical position and affords ready access. Their first egg was laid on 5 December, 1986, followed by a second egg on 9 December. As the eggs were not marked for identification and only one hatched, the exact incubation period is uncertain, but as incubation appeared to commence the day after the first egg was laid, the period was between 23 and 26 days.

1987 SEASON RECORDS

Day	*Weight* g	*Notes*
1	13.5	
2	16.5	
3	19	
4	24	
5	29	
6	34	
7	51	
8	53	
9	63.5	
10	67	
11	76	Egg tooth still visible
12	84	
13	86	
14	100	
15	105	Pin feathers visible on wings
16	115	
17	120	
18	135	
19	130	
20	140	24 g less than the chick of 1986
21	130	
22	152	
23	156	Flank pin feathers visible
24	158	
25		Weighing discontinued during holiday period
37	172	Weight gain grossly inadequate
38	169	Obvious calcium deficiency
45		Chick died despite therapy in last week

From Matthew's 1987 season records, the first egg was laid on 4 November and the second on 7 November. This season the eggs were identified but unfortunately the marks rubbed off in the nest. As only one egg hatched the exact incubation period was still not recorded but had to be between 23 and 26 days.

I find Matthews' data invaluable, for it provides us with rare information on parent reared chicks and comparisons to assist in predicting pending dietary deficiencies or parent failure. This could be applied in our aviaries or to our hand rearing. When parent rearing starts to go wrong it is always wise to start hand rearing the babies before they become too backward in development. In retrospect the second chick should have been taken for hand rearing at about day 20.

Feather Plucking

Gang Gangs are delightful, intelligent aviary birds and have been my favourite bird since I first kept them, back in my early avicultural years. But unfortunately many have a major feather plucking problem when held in aviaries.

In the past it has been suggested that much of the blame for feather plucking was related to inadequate diet and the feeding of sunflower seed. I am now convinced that diet has little or no bearing on the feather plucking problem and that it is in fact a vice which is induced by boredom, discontent or trauma. In my collection, established devoted pairs do not pluck, whereas birds held in flock conditions or with incompatible mates are more prone to pluck.

Plucking is often sparked off by stress or some trauma the bird has been subjected to. My first experience of Gang Gang plucking occurred almost 25 years ago, when I had a flock of nine Gang Gangs held in a large aviary approximately 7 m (24 ft) long, 4.5 m (15 ft) wide and 2.7 m (9 ft) high. They were all in immaculate condition when I left them on a Friday afternoon for a long weekend away. Adequate food and water was provided for the three-day period, as they would see no one until I returned. When I arrived home during the late afternoon on the following Monday, I was devastated to find the aviary floor littered with feathers and seven of the nine birds dreadfully plucked. The two birds which never touched a feather eventually became my first breeding pair and in fact the hen of that original pair reared young last season. Of the seven plucked birds five were eventually restored to full feather. Two birds continued to pluck until they died, one from the result of a crash landing from a high perch and the other died many years later from a bacterial infection.

Looking back on this experience, I feel certain that this plucking spree was a result of the trauma related to me not physically attending to them for three days.

Another more recent plucking experience occurred in my aviaries with a pair of young Gang Gangs which I had been forced to take for hand rearing due to the heat wave conditions we experienced when the chicks were at the advanced pin feather stage. (Gang Gangs are vulnerable to death from heat exhaustion from pin feather stage to fledging if left in their nesting log during temperatures in excess of 95°F.)

They were an immaculate friendly pair, seven months old and were being housed in a large holding aviary with a mixture of young parrots, when we departed for a European avicultural tour. My birds were left in the capable hands of Shirley Gardiner.

When we returned some five weeks later, Shirley informed me that three days after our departure the young Gang Gangs started to pluck and they are still plucking some three years later. I did manage to restore them to full feather for a short period, but they plucked again when placed back in an aviary. Shirley felt that the breakdown of the normal routine in which these birds had close contact with my wife and myself daily, sparked off their plucking and I must agree with her.

The best preventative and in some cases cure for feather plucking is the chewing therapy. That is, unlimited material should be provided for the birds to chew. Eucalypt and other native tree and shrub branches, and the seed pods of Banksia, Casuarina and Hakea as well as pine cones are all suitable and will all help to occupy their time and mind thus discouraging this misplaced behaviour. By the way it is probably advisable to take your Gang Gangs on holidays with you.

Mutations

I have records of the occurrence of two cinnamon Gang Gangs, one is of a field sighting and the other was a bird held in an aviary some years ago. Unfortunately this bird does not appear to have reproduced.

Hybrids

Hybrids have been recorded with Galahs and Corellas. An interesting case of hybridising in the wild was recorded recently by Mrs Appleton of Glenbrook in the Blue Mountains west of Sydney. A Gang Gang hen which is native to the area mated with a Short-billed Corella cock which is not native to the area. The youngster produced was light grey all over with a reddish suffused head, probably denoting a young male. The parents and their young fed from a local resident's feed tray, used to attract birds, for some months.

stling one day old.

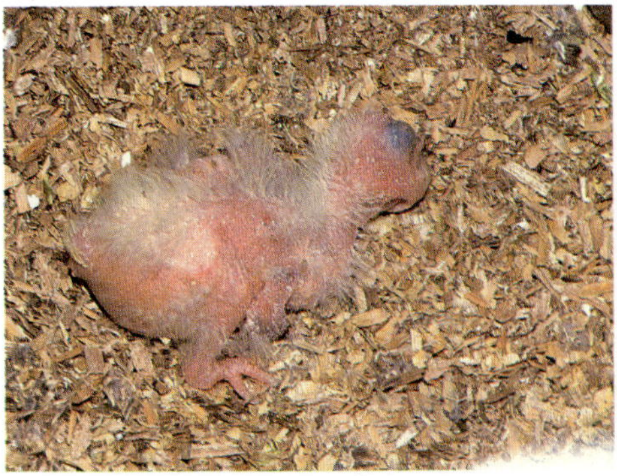

B Nestling one week old.

stling three weeks old.

D Nestling four weeks old. Note red head feathers denoting a young male.

stling five weeks old.

F Nestling seven weeks old.

Plate 18: Gang Gang Cockatoo — Nestlings.

GENUS EOLOPHUS

Plate 19. Galah *Eolophus roseicapillus*.

GALAH

(Eolophus roseicapillus)

Derivation:

Eolophus:

Eos — Greek for dawn — the red colour at dawn

lophus — Greek for the crest of birds

roseicapillus:

roseus — Latin for rosy

capillus — Latin for hair on the head, in this case meaning feathers

Description:

Length 36 cm
Weight 330 g, average
Appearance as in Plate No. 19

Classification

I agree with Forshaw in removing the Galah from the Cacatua genus and re-introducing the genus Eolophus to accommodate it. Obviously this species is distinct from all other cockatoos and should remain in this monotypic genus. However, I cannot agree with any suggestion of a distant relationship with the genus Callocephalon (Gang Gang) although there are apparently some anatomical similarities. I have kept and bred both species for over 25 years and have been unable to detect any similarities between them or between either of them and any other species of cockatoo, other than the fact that they are obviously cockatoos.

As must be expected with a species as numerous and with such an extensive range as the Galah, localised size and colour variations occur ranging from pallid forms to vivid almost red forms but such differences do not necessarily constitute subspecies.

Currently there are two recognised subspecies which are mainly distinguished by the colour of the periophthalmic eye ring. The nominate species *Eolophus roseicapillus roseicapillus* has a deep pink to red periophthalmic ring which varies in depth of colour throughout its range. (See Plate No. 20A.)

Eolophus roseicapillus assimilis, which is a western subspecies, has a white to greyish white periophthalmic ring and a noticeably fuller, longer and pinker crest. (See Plate No. 20C.) Until recently I was prepared to accept the two subspecies *E. r. roseicapillus* and *E. r. assimilis* now generally acknowledged, but after examining very closely a delightful, tiny adult pair of Galahs from the Darwin area of the Northern Territory I feel I must suggest that the reintroduction of the subspecies *Eolophus roseicapillus kuhli* could be warranted. This subspecies or race is much smaller (at least 6 cm shorter) than the other subspecies, the general colouration is more pallid and the head is small and rounded with a very high crown, particularly noticeable when viewed in profile. The birds I have been able to examine, which by the way are alive, have a more pronounced sexual difference in the periophthalmic eye ring, the males having a deep reddish pink oval eye ring and the females a less extensive dull pinkish oval eye ring. (See Plate No. 20D.)

Earliest Reports

This species was not encountered by the early voyagers to our shores, as it was then a bird mainly from the inland. It was described by Vieillot in 1817 from a specimen in the Paris Museum whose origin was listed incorrectly as the Indies. Vieillot later figured the species in the "Galerie des Oiseaux" as *Cacatua rosea*

Range, Habitat and Field Notes

Originally the Galah ranged over most of inland Australia and only came near the coast in the more arid regions of northern and western Australia. The clearing of land for agriculture and the establishment of permanent watering places both in the inland and coastal regions has enabled this already successful species to become the most successful psittacine in the world. Thirty years ago it was unknown on the coastal plains of eastern Australia and it was not until, when as a youth, I crossed the Blue Mountains west of Sydney, to get to the Western Slopes, that I first saw the Galah in the wild.

While taking some New Zealand friends on a bird watching excursion into the mountain country south-west of Sydney in the early 1960's, we observed a small flock of Galahs in grazing country near Moss Vale.This was the closest I had seen wild Galahs to the coast in New South Wales and within a few short years they had arrived on the coastal plains and an absolute population explosion occurred. This species is now common along the coast of New South Wales, Victoria and much of Queensland. It can now be seen feeding in the parks of Sydney and suburbs alongside the seagulls and feral pigeons. A breeding pair and their young regularly visit my backyard to feed on the seed in the sweepings from my aviaries.

Hindwood and McGill ("Birds of Sydney" 1958) state that large flocks of Galahs arrived in the Sydney region during a drought in 1941, but by 1944 their numbers had decreased considerably and they were then restricted to the western metropolitan area. I was too young to have observed this happening but a few years later I spent much of my spare time wandering the rural areas adjacent to the western and south-western suburbs of Sydney without ever seeing a Galah. It is my belief that this earlier attempt to extend their range to the coastal plains near Sydney failed.

So extensive has the general national population explosion been, that I believe there is very little of mainland Australia and Tasmania not yet inhabited by the Galah and I am sure that with the constant clearing of land and the cancerous spread of human habitation, it will only be a few years before the Galah will be seen over the entire Australian Continent.

The exact range of the nominate subspecies, *E. r. roseicapillus* is not known and obviously there must be some integration of the subspecies where the ranges merge. It is known to range over eastern, north-eastern, southern (except the south-west) and central Australia as well as Tasmania.

E. r. assimilis is known to range from Derby in the north, eastward to take in much of the interior of Western Australia and southward to cover all the south-west of Western Australia. The exact range is still undetermined. The western deserts appear to divide *assimilis* in the west from *roseicapillus* in the east.

The suggested subspecies *E. r. kuhli* is thought to range through the northern areas of the Northern Territory and Western Australia.

Open forest and tree studded grassland is the most preferred habitat but as the massive range suggests, it may now be seen in almost every form of habitat, its only prerequisites being a food source, water and nesting sites.

Galahs' movements vary considerably according to their habitat. In dry arid areas where the availability of food is inconsistent, they tend to be nomadic, but in fertile regions where food and water are reliable, established breeding pairs are mainly sedentary. After the breeding season flocks of juveniles and unpaired adults are formed which may number in the thousands. These flocks are totally nomadic, their movement being influenced by the availability of food.

Food

As would be expected of such a successful species, the Galahs' food requirements are as varied as the country they inhabit. They will eat seeds, nuts, fruits, roots, leaves and blossoms of various native and introduced grasses, plants, shrubs and trees together with insects and their larvae.

When travelling in arid regions of inland Australia, I have often observed large flocks feeding on bare dusty ground which does not appear capable of supporting a budgerigar let alone a flock of 500 galahs. Yet these flocks glean through the dust for minute seeds that apparently abound, as the birds may be seen combing the same area day after day.

In grain growing areas the galah causes a certain amount of crop damage, but I am sure that excessive damage is confined to certain areas in particular seasons, just as fire, flood, hail, grasshopper plagues, etc., are. The continual national statistics of near record grain harvests do not reflect the supposed general massive crop damage carried out by the birds. I am none the less very sympathetic towards the individual farmer who may experience significant losses.

Breeding in the Wild

The breeding season varies according to climatic and weather conditions but usually takes place during winter, spring and summer in most parts of Australia. In the far north breeding occurs after the wet season, which usually means nesting commences in late autumn or early winter. Breeding in the arid inland is related totally to rainfall and in dry seasons often does not take place. Generally speaking the further south the habitat the later the breeding season.

Galahs are usually single brooded but in good seasons they may nest twice. The nesting site is usually a hollow in a dead or living tree, often a Eucalypt, at heights varying from 2 m (6 ft 6 in.) to 20 m (65 ft) or even higher at times. Nest holes in rocky outcops and cliffs have been recorded. Both the male and female prepare the nesting site, spending considerable time chewing away bark and rotting wood for a large area around the entrance hole. Nests used for a number of seasons become as smooth as glass, around and up to a metre below the entrance hole. The talcum powder-like dust deposited from the bird's feathers becomes impregnated into the timber making it very slippery. It has been suggested that this habit is a defence mechanism against predators such as lizards and snakes.Usually three or four white oval eggs (occasionally two or five) are laid on a bed of Eucalyptus leaves prepared by both birds. Additional fresh leaves are usually added to the nest during the incubation period, possibly to increase the humidity. Incubation is carried out by both the male and female and when the chicks hatch both parents feed them.

When parent birds approach the nesting site they do so cautiously, although often noisily, never flying directly to the nest and usually taking many minutes to casually work their way to the entrance, and when perched on the mouth of the nest, will rest inconspicuously for some time before quietly slipping away, mostly rear first into the nest. After the young fledge they are taken to a communal nursery, usually in a protected clump of trees, where they are cared for by their parents, who apparently believe there is safety in numbers. These nurseries are readily identified by the constant whining, crying noises emitted by the fledglings. Such a nursery was observed in a dense stand of shade trees adjoining Musgrave Station Homestead on Cape York Peninsula during a field trip in July 1983. The noise from this nursery, which appeared to contain a hundred or so fledglings, was deafening, due to their continual begging cries for food and the screeching of the parent birds which were constantly coming and going.

General

This cockatoo is a likeable bird which enjoys life to the full, apparently gaining much pleasure from its everyday activities. There must be but a few Australians who have an interest in birds, that have not enjoyed the spectacle of a flock of galahs flying in wild erratic unison. For those readers who have not witnessed these aerobatic flights, I will endeavour to describe them. A flock will suddenly appear, screeching, twisting, turning, zooming just above the ground, then soaring up among the treetops only to dive low again to buzz the motor vehicles speeding along a country road. The whole event is carried out with a wild abandonment and the birds themselves seem to ooze with the pleasure of living. Unfortunately such flights often result in death due to collisions with motor vehicles.

Aviculture

There are many avicultural references relating to Galahs both in England and Europe during the 1870's and the first assisted breeding in which the eggs were incubated and hatched under a bantam hen and the young hand reared, because the female had died during incubation is credited to Mr Gedney (England). This success occurred some time prior to the published account in 1876 (Foreign Cage Birds Vol. 1, page 146). The first recorded parent reared breeding was by the Marquis of Tavistock (later to become the Duke of Bedford) at Woburn, England, in 1910.

Unfortunately, as is often the case, all early Australian breeding successes went unrecorded. Lendon mentions four chicks being hatched by an unnamed lady in 1937 without giving further details. The first recorded breeding appears to be from the parrot records at Adelaide Zoo which states Galahs were bred in a large mixed collection within the zoo in 1940.

This cockatoo is possibly the most commonly kept caged parrot, except for the budgerigar, in Australia and the cage bird trade in galahs flourishes. Wild caught Galahs, either adult or juvenile, are totally unsuited to life as cage birds. The organised collection of nestlings, particularly from areas where crop damage is prevalent, would be a far more humane method of controlling populations as well as providing suitable birds for the cage bird trade. Hand raised Galahs make pleasant and desireable pets.

Galahs are rarely kept in Australian aviaries although the constant appearance of wild caught mutations during the last decade or so has prompted many aviculturists to make aviary space available to them.

Sexing

Adult Galahs are readily sexed; the female has a coppery red iris and the male has a dark brown iris. (See Plates Nos. 20A and 20B.) All immatures have a light brown iris but young males usually have a larger head and mandible than females.

Display

This cockatoo has a strong pair bond, in fact field reports suggest that they mate for life. Established pairs enjoy sitting together for hours engaged in mutual preening. Their display in aviaries consists of the male excitedly approaching the female with his crest erect, often with wings spread slightly and head wobbling from side to side. His stance is erect as he advances along the perch uttering a soft clicking note. If the female is receptive she will greet him with her crest erect and much bobbing up and down.

Nests

Galahs usually prefer vertical hollow logs with natural entrance spouts as nesting sites. I have had them accept logs with internal diameters ranging from 30 cm (12 in.) to 60 cm (24 in.) and depths of 45 cm (18 in.) to 1.5 m (5 ft) below the entrance hole. In captivity as in the wild much

ninate subspecies *E. r. roseicapillus* — Male. e whiter crest and pink eye ring.

B *E. r. roseicapillus* — Female. Note copper coloured iris.

. *assimilis*. Note white eye ring and fuller, k crest.

D *E. r. kuhli*. Note pallid colouring, small head and elliptical reddish eye ring.

Plate 20: Galah subspecies.

A Nestlings one day old.

B Nestlings three weeks old.

C Nestlings four weeks old.

D Nestlings five weeks old.

E Nestlings six weeks old.

Plate 21: Galah — Nestlings.

time is spent chewing and smoothing the entrance hole and adjacent areas of the log. The nest chamber is lined, or should I say filled, often to a depth of 15 cm (6 in.) or more with Eucalyptus leaves or any herbage available such as grass, weeds or silverbeet, etc. Stones and bones are also often placed in the nest.

Nesting and Hatching

In my aviaries three to four eggs is the usual clutch, laid at intervals of two or three days. (usually two days). Incubation is generally carried out by the female for the first day or two and is then shared with the male. The male usually incubates during most daylight hours with the female incubating at night. Often when the nest is left unattended for a short period, the eggs are covered or partially covered with herbage.

Hatching occurs usually after an average of 24 days, although this is variable. I have recorded incubation periods of 23, 24 and 25 days' duration. The chicks hatch with the slightest trace of a whitish down which gradually disappears.The eyes start to open and the pin feather development is obvious at two weeks and the chicks are fully feathered at six weeks of age. Fledging usually occurs about seven weeks after hatching. The young leave the nest as smaller, duller replicas of their parents with a light brown iris and pale grey eye ring.

The young are fed by both parents for some weeks after leaving the nest and can be constantly heard begging for food. Immature birds eye colour changes gradually from five to six months of age onwards and is similar to the adults in the second year. I have not found them to be sexually mature and capable of breeding until they are three years old. Galahs appear to have a lifespan similar to that of humans and are capable of breeding up to forty years of age.

Again I must stress that obesity is a major problem with Galahs, particulary after they are six or seven years old and unless they are kept in a lean condition they will not breed. It is essential to feed a nonfattening, low calorie diet. This consists of a handful of seed per bird per day as suggested in the chapter on Diets.

A couple of years ago I was visited by Mr Bernard Roer of Phoenix Arizona, USA, who is a highly experienced professional aviculturist. He has bred Galahs in his aviaries for at least three decades and would have to be regarded as one of the world's most experienced breeders of Galahs. At the time of his visit my breeding pairs of Galahs had not nested for three seasons and were obviously overweight. At first I had blamed the drought conditions, but the drought broke and still the birds did not breed. By the time Bernard visited me the birds had been on an all-millet diet for over 12 months, but still they were fat. Bernard noticed the obesity problem immediately and said that he had experienced the same problem some years before in the USA. He eventually solved the obesity syndrome by feeding each pair of Galahs only one small handful of mixed sprouted and dry seed per day, together with plenty of green food. I placed my Galahs on crash diet, moved them from 5 m to 7 m aviaries to increase their exercise and fifteen months later had baby Galahs in the nest again after four seasons of failure.

Mutations

I have recorded more mutations in Galahs than in any other species of Cockatoo. They are:

Albino (sex linked) (See Plate No. 22A)

Dilute (recessive) (See Plate No. 22C)

Cinnamon (two distinct simple recessive forms and one sex linked form) (See Plate No. 22B)

Silver (another form of cinnamon?) (See Plate No. 22E)

Off-white, lacking both pink and grey

Grey and White — a mutant where the pink is replaced by off-white, similar to the blue mutation of green coloured birds, almost certainly a simple recessive inheritance. (See Plate No. 22D)

Most of these mutations are currently being worked on by Australian Aviculturists but unfortunately I cannot say that any are established in aviaries. I have bred a recessive cinnamon and an albino in my aviaries.

Hybrids

Hybrids have been recorded with the Gang Gang, Sulphur-crested Cockatoo, Major Mitchell, Short-billed Corella, Long-billed Corella and the Lesser Sulphur-crested Cockatoo. Forshaw records wild hybrids with the Major Mitchell and the Short-billed Corella.

ɔino. Note white is snow white, pink feet, red
:s.

B Cinnamon. Note fawn flight feathers, dark eyes in male and pink feet.

ıute. Note off-white colouring, normal eyes
ı feet.

D Grey and White Galah. Note pink replaced by white, grey intensified, eyes and feet normal.

E Silver. Note. A form of recessive cinnamon.

Plate 22: Galah mutations.

GENUS CACATUA

CACATUA

The genus Cacatua is one of convenience, in which all species of white or near white cockatoos have been placed together, regardless of their many diversities. Until recently the Galah was also included in this genus, but with new information at hand Forshaw wisely saw fit to reintroduce the genus Eolophus, of which the Galah is the sole member.

Although I am prepared to adopt and accept, with some reservations, the current species membership of the genus Cacatua, I cannot accept some of the subspecies structure currently in vogue.

My personal opinion is that many of the species included in this genus are diverse enough to warrant separate genus status. Many years ago, the renowned English aviculturist E. N. J. Vane wrote the article "Some Observations on the Cacatuinae — The Cockatoos", which was duly printed in the Avicultural Magazine, Volume 65, 1959, pages 9 to 16. He explained and suggested an acceptable and comprehensible taxonomic restructure of the genus Cacatua involving its expansion into four genera. Recent developments, including the reclassification of the Galah, have made parts of Vanes' suggested restructure inapplicable. By adopting his basic ideas and making what I believe to be the alterations necessary for a working alternative to the genus Cacatua, I propose the following four genera.

Cacatua — to include the: Great White Cockatoo, *C. alba*; Moluccan, *C. moluccensis*; Blue-eyed Cockatoo, *C. ophthalmica*; Greater Sulphur-crested Cockatoo, *C. galerita*; Lesser Sulphur-crested Cockatoo, *C. sulphurea*; with the only Australian member being the Greater Sulphur-crested Cockatoo, *C. galerita.*

Lophochroa — to be a monotypic genus with the Major Mitchell Cockatoo, *L. leadbeateri*, the sole member.

Eolophus — in which Vane included Ducorpsius, but which has since become a monotypic genus with the Galah, *E. roseicapillus*, as the sole member.

Ducorpsius — to include the Australian Corellas in three distinct species: Short-billed Corella, *D. sanguinea*; Western Long-billed Corella, *D. pastinator*; Eastern Long-billed Corella, *D. tenuirostris*; as well as the exotic members: Ducorps Cockatoo, *D. ducorpsi* (nominate subspecies); Goffins Cockatoo, *D. goffini*; Red-vented Cockatoo, *D. haematuropygia*, etc.

Vane suggested the Long-billed Corellas be installed into a separate genus, Licmetis, which I feel would be unwarranted with the introduction of the genus Ducorpsius.

SULPHUR-CRESTED COCKATOO

Plate 23: Sulphur-crested Cockatoo *Cacatua galerita.*

SULPHUR-CRESTED COCKATOO

(Cacatua galerita)

Derivation:

Cacatua — Latinised version of *kakatua* which is Malay for large parrot

galerita — from *galeritus* which is Latin for crested

Description:

Length 50 cm average

Weight 870 g average; females are generally heavier than males

Appearance as in Plate No. 23

Classification

The Sulphur-crested Cockatoo has an extensive range through northern and eastern Australia, West Irian, Papua New Guinea and numerous adjacent islands and island groups, as well as many of the Indonesian Islands.

There are four subspecies recognised by Forshaw in his second edition of "Parrots of the World".

Cacatua galerita galerita is the nominate subspecies from the eastern half of Australia and is distinguished by a white periophthalmic eye ring and only slight yellow suffusion on the ear coverts. (See Plate No. 24A.)

Cacatua galerita fitzroyi is from the north-western section of the Australian range and meets *C. g. galerita* in the Gulf of Carpentaria. This subspecies is distinguished by a pale blue periophthalmic eye ring, yellower ear coverts and longer crest feathers. (See Plate No. 24B.)

Cacatua galerita triton ranges through Papua New Guinea, West Irian and numerous adjacent islands and island groups as well as some of the Indonesian islands. The distinguishing features of this subspecies are broader, more rounded crest feathers, a blue periophthalmic eye ring and it is smaller than the Australian subspecies.

Cacatua galerita eleonora from the Aru Islands of Indonesia is smaller than *C. g. triton* and has a much smaller beak.

Note the Lesser Sulphur-crested Cockatoos do not range into Australia and are classified as the separate species *Cacatua sulphurea.*

Earliest Reports

This species was first considered similar if not identical to the previously described White Cockatoo of the Moluccas. Thus the first illustration of *C. g. galerita* in White's "Journal of a Voyage to New South Wales" dated 29 December, 1789, is under the name of Crested Cockatoo. Then in the text on page 237, the name *Psittacus cristatus* (the scientific name then used for the White Cockatoo) is given to the Sulphur-crested Cockatoo, noting that it is considered to be only a variation of the previously described White Cockatoo.

Latham described and named this species in his "Index to Ornithology" Volume 1, page 109 in 1790 as *Psittacus galeritus*.

Probably the best of the early field notes are those by George Caley, published by Vigors and Horsfield in the "Transactions of the Linnean Society", London, Volume 15, page 268, 1827.

Range, Habitat and Field Notes

Sulphur-crested Cockatoos range over northern, eastern and south-eastern Australia as well as Tasmania, with an isolated and growing introduced population in the Perth region of Western Australia.

The nominate subspecies, *C. g. galerita*, ranges eastward from the southernmost section of the Gulf of Carpentaria and the Mount Isa-Cloncurry region of Queensland, where it appears to integrate with *C. g. fitzroyi* and then extends over the entirety of eastern Australia as far west approximately as the junction of the Thompson and Barcoo Rivers of central western Queensland, down to the Menindee Lakes area of western New South Wales, and as far west as Spencers Gulf in South Australia, as well as Tasmania.

C. g. fitzroyi ranges from the Gulf of Carpentaria, Cloncurry, Mount Isa regions of Queensland, westward across the whole of northern Australia as far south as Tennant Creek in the Northern Territory to Broome and the Fitzroy River in Western Australia.

Introduced populations of *C. g. galerita* are well established in an extensive region around and west of Perth in Western Australia, as well as on the east coast of the north island of New Zealand.

This species is becoming more numerous in the populous areas within its range, particularly around Sydney where it is now common throughout the whole of suburbia.

My first experience of this species in the field was during the 1940's when as a child I would often go "rabbiting" in the Cecil Park area, south-west of Sydney, where a small, nervous population of Sulphur-crested Cockatoos lived and bred. This group was difficult to approach and they always placed a sentinel in a high overlooking tree, when they were feeding on the ground, who was replaced periodically by another member of the flock so all could feed. Should anyone approach them, particularly with an unconcealed firearm, they would noisily vacate the area and would not return until the intruder had gone. My early impression of this species was of an intelligent, wary, perceptive bird. Now some forty-odd years later I find it hard to believe that this is the same species that now visits my back garden in the south-western suburbs of Sydney, not 16 kilometres from where I first saw this species in the wild. They delight in disrupting and panicking the birds in my aviaries, chopping huge pieces off my trees and shrubs, chewing at electric wires, television aerials and the timber on out-buildings and refusing to be driven away. What has happened to this shy, wary species? Surely this change has happened too quickly to be classified as evolution. Perhaps it is just the unfortunate effect city dwelling has on all living things.

Property damage caused by Sulphur-crested Cockatoos is becoming increasingly more common throughout the Sydney metropolitan area, particularly in areas adjacent to national parks or where misguided but well meaning people provide food trays for the Cockatoos.

The panic created in my aviaries by a Sulphur-crested Cockatoo flying over is identical to that caused by a bird of prey, yet I doubt that the Cockatoo is being confused with a bird of prey, being totally white and having a completely different pattern of flight. There is no apparent reason for the smaller psittacines to fear Sulphur-crests, yet they do and not just because they are a large bird reminiscent of a bird of prey, for Galahs are totally ignored, even when flying within a metre or two of the tops of my aviaries.

My observations of this species in the field suggest that it is far more numerous in the southern half of its range than in the north. Large flocks of hundreds of birds are often encountered in the southern areas, whereas I have only observed pairs or small parties in the northern sectors.

The sight of many hectares of sunlit, bright green pasture, with the colour accentuated by the snow-white spotting of several hundred Sulphur-crested Cockatoos foraging for food is something to behold, but unfortunately is so seldom appreciated in Australia by Australians. What is an even greater sight is that of that same flock of Cockatoos, when startled by the alarm calls of their sentries, as they rise in unison like hundreds of pieces of white paper fluttering in the breeze.

I have observed the subspecies *C. g. fitzroyi* many times in the field but have been unable to detect the difference between it and *C. g. galerita.* As is the case with many subspecies, close scrutiny in the hand is necessary for identification. My impression in the field of *C. g. fitzroyi* is that it is far less numerous and more scattered throughout its range than *C. g. galerita* and not inclined to congregate in large flocks.

Although *C. g. galerita* is thought to inhabit the whole of Cape York Peninsula, my field observations, as well as avicultural information presented to me, suggest that the Sulphur-crest inhabiting the Iron Range rainforests of this region is a smaller, shorter tailed cockatoo with more pronounced yellow ear coverts. They also appear to have a pale blue periophthalmic eye ring. I found this difficult to detect in the field, but aviary specimens known to have originated in the area have these features. They also have a distinctive call and appear to be of a solitary disposition with only single birds or pairs observed. A more intensive study of the *C. galerita* from the Iron range is warranted to determine if this Cockatoo is in fact *C. g. galerita* or more closely allied to *C. g. triton* or *C. g. fitzroyi.*

This species frequents a wide range of habitats from rainforest and mangrove swamps to Eucalypt forests, open forest, farmland, mallee and treelined watercourses. Generally, this cockatoo feeds in open country and returns to stands of trees to rest during the heat of the day and to roost at night. Individual groups usually use the same roosting area all year round, providing food and water remains available within a reasonable flying distance.

Sulphur-crested Cockatoo populations are reasonably stationary with mainly localised movements being observed, although in some areas seasonal movements are normal and on occasions weather conditions or food availability will promote a large influx or exodus in particular regions.

The diet varies considerably and includes vegetable matter, blossom, seeds of grasses, trees and shrubs, as well as nuts, fruits, roots and various insect forms, together with the seeds and fruits of introduced flora including Hawthorn and Pine seeds. Unfortunately,. their liking for ripening grain causes considerable damage in farming areas and has been and still is the reason for their constant persecution in grain growing areas.

In recent years there has been considerable hysteria and debate between farmer groups, bird trappers and would be bird exporters on one side and conservation and animal welfare groups on the other, together with numerous other divided sections of the community such as aviculturists. The first group feel that all excess cockatoos should be trapped and exported so as to rid the agricultural areas of a pest species and earn export dollars at the same time, not to mention making wealthy men of the bird trappers and exporters.

Of course the second group oppose the trapping and exploitation of pest species for reasons varying from morality and cruelty to a general concern that such exploitation of pest species may gradually extend legally or illegally into less abundant species.

Aviculturists appear to have a wide range of opinions on the subject.

I believe that crop damage caused by cockatoos is no different now than it has been in the past. There has always been varying degrees of crop damage in certain areas in particular years. This is related to fluctuations of cockatoo populations due to good seasons stimulating exceptional breeding conditions, or to mass movements promoted by adverse seasonal conditions in other areas. Such situations of major crop damage should be categorised as natural disasters along with fire, flood, grasshopper plagues, mouse plagues, etc., and handled accordingly at the particular time and place by the authorities concerned, but in the case of cockatoos, protection should prevail when they are not in excessive numbers.

Our wildlife authorities must police and eradicate the practice of poisoning food and water supplies as a means of controlling pest species, due to the untold damage inflicted on protected species by this all too frequently used barbaric method of control.

All eradication and other methods of control must be supervised by wildlife officers, and carried out in a responsible and humane manner if the acceptance and support of the general public is to be gained by the farmer groups and wildlife authorities alike.

The flight of this species consists of several rapid, short wingbeats, which vary in number, followed by a brief moment of gliding flight. A period of flight is usually completed by gliding in for a landing. During such gliding flight, the short, almost square tail and rounded wing tips symbolic of this species are viewed to the best advantage.

Breeding in the Wild

The usual nesting site is a large hollow limb or a hollow in the trunk of a tree, usually at least 10 m (30 ft) and often much higher above the ground. The nests I observed as a boy were always hollows in large dead trees at least 15 m (49 ft) high. Most other nest sites I have seen have been in large Eucalypts, usually close to or bordering water courses. The hollows formed where a large limb has broken off. In north-west Queensland I have observed this species nesting in River Gums (Eucalypts) bordering dry creek beds on many occasions.

Both male and female prepare the nest site by chewing around the entrance and the inside of the hollow and sometimes carry out considerable excavation to reach the desired depth. The preparations are often interrupted by courtship displays and mutual preening.

Breeding has been recorded as early as May in the north and as late as January in the south of Australia. Generally speaking the breeding season commences in late autumn or winter in the far north of the range and occurs progressively later the further south the species is found.

Two or three white oval eggs are laid on a base of rotten wood dirt. Incubation is shared by the male and female, who are secretive and silent whenever in the vicinity of their nest, always approaching cautiously and slipping quietly into the hollow and leaving in a similar manner. Reflecting on the thousands of times I have observed this species, the only occasions I have found them to be silent is during nesting operations or when they are feeding.

The chicks hatch covered by a yellow down, are brooded closely by the parents for two or three weeks and fledge when about 10 weeks old. Both parents feed the young in the nest and for several weeks after fledging, when they can often be seen, perched high in a tree, crying and begging in such a forlorn way that their parents eventually relent and feed them.

The formation of large flocks after nesting is not as evident in the northern range as in the south.

Aviculture

In Australia this cockatoo has historically been kept as a pet cage bird and unfortunately seldom as an avicultural subject, due mainly to its abundance and the constant availability within the pet trade of legally trapped birds.

Fortunately this species has received far more attention from overseas aviculturists, both in the past and currently, than it has at home.

The first recorded breeding was by Mme Eugenie Delaitre of Saint Eugene, near Algiers, who housed a pair in a cage 80 cm (32 in.) by 50 cm (20 in.). They successfully bred in these cramped conditions, rearing one young in 1879 and two in 1880. No details are given of the nesting facilities provided. In 1881 Mme Delaitre moved her birds to Nice, France, where they reared another youngster in that year.

Then in Germany, Ernst Dulitz of Friedrichschagen, near Berlin, reared two young of this species from a pair housed in a garden aviary during 1883.

Early English successes with this species were from birds kept at liberty or semi-liberty. The first recorded successes were Charles Buxton's liberty breeding at Northreppes in Surrey, prior to being reported in 1892.

As usual, early Australian avicultural records are non-existent for this species, although it has undoubtedly been bred many times this century. Over the last decade many cockatoo specialists have bred this species in Australia but none have bothered to record their breedings because the species is so common.

My own pair, which are hand reared and over 20 years old, regularly hatch chicks but refuse to rear them to independence.

An interesting breeding of this cockatoo occurred at Kemps Creek, which is an outer rural suburb of Sydney, on Glenda Bain's property. Glenda's pet hen Sulphur-crest is unable to fly, so she is allowed to wander the acreage at will. During the winter of 1985, two wild males, who may have originally been escapees from a local aviary, started to visit the hen and soon a pair bond was formed with the most dominant male.

As spring approached this dominant male selected a hollow in a nearby Eucalypt and tried everything he knew to entice the hen up the tree to inspect the hollow, but try as he may, the flightless female found it impossible to ascend the smooth barked tree. A rough ladder was constructed and lent against the tree to provide access to the proposed nesting site for the hen. She soon made the climb to the hollow and inspected it excitedly, but her obvious delight was soon to become anguish when the weather changed and gale force winds blew in. The hen clung to the broken limb at the entrance of the hollow for three days, until the wind dropped sufficiently for her to make the climb safely to the ground.

Once on the ground, she lost all interest in climbing trees and soon commenced to prepare a nest site in a wooden chicken nest box in the security of a fowl shed. The male, unable to entice the hen back up the tree, decided to accept her chosen site. Two eggs were laid in early October and were incubated by the tame hen as well as the wild male, who came and went as he pleased. Both eggs were hatched in early November and the youngsters reared to independence by both parents. When the young birds were deemed capable of fending for themselves they were allowed to join the local wild population, but both still return occasionally for a brief visit.

In 1986 the pair nested again under similar circumstances except that the chosen nest site was an old metal feed hopper and only one chick was reared. At the time of writing, September 1987, nesting activities are commencing again.

Sexing

There has always been considerable confusion with regards to the sexing of this cockatoo. Forshaw states incorrectly that males have a blackish brown iris and that females have a reddish brown iris. All birds with a reddish brown iris are females and all males have a blackish brown iris, but many mature females also have a blackish brown iris.

Whether the occurrence of a reddish brown iris in some females is a locality, racial or age factor is still undetermined, but known egg laying females, with a blackish brown iris, that are in excess of 20 years old have been reported.

Immature birds are extremely difficult to sex.

Display

The male advances towards the female in an excited strutting manner, with his crest erected and tail spread wide, while nodding and flinging his head from side to side and emitting a low clucking call.

Nests

Various hollow logs and nest boxes have been used for this species, but hollow logs of Australian hardwoods, such as Eucalypts, are best suited to withstand the destructive beak of this cockatoo. Suspended hollow logs about 1 m (39 in.) long and approximately 35 cm (14 in.) internal diameter, in either a vertical or inclined position, are often accepted. If one type of nest is not to a pair's liking another should be tried.

ɔminate subspecies, *C. g. galerita*. Note white e ring.

B *C. g. fitzroyi*. Note long crest and pale blue eye ring.

ːwly hatched chick.

D Nestling three days old.

ːstling three weeks old.

F Nestling six weeks old.

Plate 24: Sulphur-crested Cockatoo — Subspecies and Nestlings.

My pair of Sulphur-crests nest in a vertical log 1.25 cm (4 ft) high and 37 cm (15 in.) in diameter, with the entrance straight down the top. It stands on the concrete floor of the sheltered section of the aviary. The log is filled to within 0.5 m (20 in.) of the top with a mixture of dirt and wood shavings. When breeding activities commence 30 cm (12 in.) or more of the filling is usually excavated before laying commences.

Should a nest box be considered for this species it would need to be approximately 1 m (39 in.) long by a minimum of 30 cm (12 in.) square with a large entrance hole near the top. It should be constructed of at least 2.5 cm (1 in.) hardwood planks and hung vertically or partially inclined.

Nesting and Hatching

Breeding usually commences with both male and female attending the nest hollow amid much displaying and mutual preening. The pair usually spend much time chewing around the top of the nesting log and often excavate some of the nest filling.

In my aviaries this species has laid eggs from early September to late December in clutches of two and three eggs laid at two and three-day intervals.

Incubation usually commences with the laying of the first egg and is shared by the male and female. The hen normally incubates for the first day or two, then the male incubates during most of the daylight hours and the female incubates throughout the night. I have recorded incubation periods of 25 days, 26 days and 27 days.

The following egg laying and hatching dates were recorded in my aviaries during 1986.

Clutch 1: 11 September, 1986, laid first egg; 13 September, 1986, laid second egg. Incubation commenced as soon as first egg was laid.

6 October, 1986: first egg hatched after a 25-day incubation period. 9 October, 1986: second egg hatched after a 26-day incubation period.

Clutch 2: 26 October, 1986, laid first egg; 29 October, 1986, laid second egg; 1 November, 1986, laid third egg. Incubation commenced with the laying of the first egg.

21 November, 1986: first egg hatched after a 26-day incubation period. 23 November, 1986: second egg hatched after a 25-day incubation period. 26 November, 1986: third egg hatched after a 25-day incubation period.

Clutch Three: 14 December, 1986: laid first egg; 17 December, 1986: laid second egg. Incubation commenced the day after the first egg was laid.

11 January, 1987: first egg hatched after a 27-day incubation period. 13 January, 1987: second egg hatched after a 27-day incubation period.

The chicks hatched with a yellow down and a horn coloured beak. At four or five days old the chicks can stand and lift their heads, then by two weeks the eyes have opened and white pin feathers are starting to appear. When three weeks old the upper mandible has commenced to darken and by six weeks both mandibles have changed colour to varying shades of brown and the yellow crest feathers are well developed. At 10 weeks the babies are fully feathered, the crest well developed and the mandibles are dark grey. Fledging periods of from nine to 12 weeks have been recorded.

Both parents feed the young in the nest and continue to do so for several weeks after fledging. The youngsters can be safely removed from the parents three months after leaving the nest. Of course, if any serious aggression towards the young by the parents is observed, they should be removed immediately. If they are not feeding themselves adequately then supplementary feeding may be necessary.

Mutations and Hybrids

I have recorded a Lutino Sulphur-crested Cockatoo in Victoria, which is similar to the normal Sulphur-crested Cockatoo in colour, with a pale horn coloured beak, flesh coloured feet and red eyes.

Hybrids have been bred from this species and the Major Mitchell, Galah, Short-billed Corella and the Eastern Long-billed Corella.

MAJOR MITCHELL COCKATOO

Plate 25: Major Mitchell Cockatoo *Cacatua leadbeateri.*

MAJOR MITCHELL COCKATOO

(Cacatua leadbeateri)

Derivation:

Cacatua — Latin from the Malay word *kakatua* for large parrot

leadbeateri — from the English naturalist Leadbeater

Description:

Length 40 cm
Weight 400 g average
Appearance as in Plate No. 25

Classification

This cacatua's appearance suggests a close relationship with *Cacatua galerita* (Sulphur-crested Cockatoo) but some aspects of their natural history sets them apart. Major Mitchells have naked hatchlings in contrast to yellow down covered hatchlings for Sulphur-crested Cockatoos. Major Mitchells are more obviously sexually dimorphic than Sulphur-crested Cockatoos. The call of the Major Mitchell is unlike any other cacatua.

Major Mitchells have an extensive range in which two subspecies are recognised. *Cacatua leadbeateri leadbeateri* is the nominate subspecies from the eastern and central parts of the range. *Cacatua leadbeateri mollis* is from the south-western parts of the range and is distinguished from the nominate subspecies by its darker red crest which has only a slight, if any, yellow band. (See Plate No. 26A). The exact extent of the range of this subspecies is not known. Another suggested subspecies, the pallid *Cacatua leadbeateri superflua* from South Australia, is not accepted by current authorities.

Earliest Reports

This cockatoo was not recorded until 1830, due no doubt to the mainly inland and arid nature of its range. It was described and named by Vigors in 1831. The first illustration appears to be in Mitchell's "Three Expeditions in Eastern Australia", Vol. 2, Plate 50, 1838.

Range, Habitat and Field Notes

The Major Mitchell ranges over most of inland Australia with the exception of the northern and north-eastern areas. It has been recorded as far reaching as the vicinity of Miles, Queensland, in the east; the Newcastle Waters area, Northern Territory, in the north; the St Arnaud District, Victoria, and the Great Australian Bight in the south; and to the Western Australian coast at the Geraldton-Northampton regions. This extensive range does not appear to have diminished during the last century, but their numbers have diminished dramatically in many areas. The clearing of the dry scrublands and woodlands for grazing purposes plus the trapping and taking of nestlings for the pet trade have greatly affected this cockatoo's population.

My good friend Marie Payne (*née* Lemon) who was raised in the small south-west Queensland town of Wyandra, told me that fifty-odd years ago (sorry, Marie) the Major Mitchell or Cocklerina, as they were known locally, were the most common cockatoo to that area at that time. They were very plentiful and because of their habit of perching in lines along the top strand of a wire fence they were easy targets. Many could be brought down with a single shotgun cartridge to be used for bait in the fish and crayfish traps set regularly in the local river.

During the last 25 years I have often passed through this country from Cunnamulla to Wyandra and on to Charleville and have usually seen a pair or two and sometimes a group of five or six birds. In May 1986, whilst passing through on a trip to the Gulf Country, I saw the most beautiful flock of 60 or more birds. One can only hope that the Major Mitchell is making a comeback, in this area at least.

It is now six months since I wrote the initial draft of this chapter and now, prompted by a recent and unforgettable experience in the field with this species, I feel compelled to record the following events.

My friend Col Duane and I were granted a few days leave of absence by those who must be obeyed, for a quick bird watching trip to south-west Queensland. Col was a new visitor to this part of the country and keen to see all the psittacines native to the area.

On the morning of 12 July, 1988, we were cruising an outback road in search of bird species and had been rewarded with good sightings of Yellow-vented Blue Bonnets, Mallee Ringnecks, Crimson Wings, Cockatiels, Galahs and Short-billed Corellas. Then two squat, whitish cockatoos were spotted perched in a small dead tree. Observation with binoculars proved them to be Major Mitchells, then further scanning revealed several more feeding on the ground in the stunted scrub studded, normally arid, grassland which had recently received 5 in. (125 mm) of rain.

We approached cautiously on foot, as Major Mitchells, although not a nervous cockatoo, usually discourage close observation by constantly moving away from would be observers. As we came closer we realised we were on the edge of a loose flock of well in excess of 50 birds. With the background cover of a few trees, we were observing one group of 20 or so Majors busily foraging 100 metres away when this most unusual event occurred. The birds obviously aware of our presence arose from the ground uttering their usual alarm call, then instead of moving away as would be expected, flew straight towards us and alighted in three strategically situated dead trees no more than 20 metres away, to observe us. As if this behaviour was not strange enough, individuals from each group set out on reconnaissance flights from their respective trees, to circle a few metres above our heads, so as to gain a closer look, then return to their perching places. One adult hen carried out this procedure three times in a few minutes, calling excitedly whilst on each circuit. Although this unique situation of the observer becoming the observed is a little disconcerting for a human, it must be classed as one of the great bird experiences of my life. The only other occasion I have had wild psittacines come to view me was yet another unforgettable experience with Keas in New Zealand.

As we returned to our vehicle after being observed for 15 or 20 minutes, Major Mitchells could be seen moving through the patchy scrub all around us. Col, who was ecstatic after his first and what will probably be his best ever sighting of Majors, suggested that we were in the midst of an army of Major Mitchells, but after contemplation we decided that perhaps it was only a platoon.

That afternoon, while boiling the billy for lunch on the bank of the Warrego River, which skirts the western side of the township of Cunnumulla, we noticed a lone male Major Mitchell behaving strangely in a Eucalypt growing on the river's edge. After watching for some time we were rewarded by the emergence of his mate from a hollow in the main trunk of the tree, about 12 metres from the ground. After this initial appearance from the hollow, both birds went in and out continually for an hour or more, engaged in preparing the nest site. The female was by far the most industrious and spent the majority of time in the hollow.

Then this pair confirmed my belief that Major Mitchells are the dumb blondes of the cockatoo world by embarking on several noisy flights across the river. They encircled the local school grounds, calling constantly, as much as to say, "Look, kids, this is where we are going to breed," before returning directly to their selected hollow. This pair actually invited the local children to rob their nest.

Some years ago I sighted and photographed a small flock of the pallid South Australian race on the eastern side of the Nullarbor Plain which were feeding in dry, sparse, low scrub country.

The Major Mitchell cockatoo is often sedentary in fertile areas with permanent water, but in the harsh arid country which comprises the majority of their range they are usually nomadic, being forced to wander in search of food and water.

Food

Their diet consists of seeds, berries, nuts, fruits and roots of native and introduced flora as well as some insects. One of their most favoured foods is the seed of the Paddymelon. In fact, most of my sightings of this cockatoo in the wild have been of the birds feeding on Paddymelons growing on the sides of country roads. To reach the seeds of this melon the birds must chew through an incredibly bitter tasting flesh, but this does not deter them in any way.

Breeding in the Wild

As with most of the Australian Cockatoos with extensive ranges, breeding occurs earlier in the north and later in the southern areas. Nesting has been recorded as early as May in the Northern Territory, during the winter months in Queensland, then through spring and as late as December in the southern parts of the range. The nesting sight is usually a hollow limb or hole in a tree, often in river gums bordering water courses. They may range in height from 3 m (10 ft) to 20 m (65 ft) or more above the ground. Both the male and female prepare the nesting chamber and a clutch of two to four white oval eggs are laid on a bed of chewed up rotten wood and sometimes strips of bark taken from around the entrance to the nest. Incubation is carried out by both the male and female. When the young leave the nest they remain with the parents, possibly until the approach of the next breeding season, which would account for the sightings of small parties or family groups so often recorded.

The Major Mitchell's call notes differ from other members of the Cacatua, being higher with a more musical quality than the harsh raucous calls of the other members of the genus.

Their flight consists of a series of wingbeats followed by a short duration of gliding. I have never observed this cockatoo fly direct long distances, always preferring to break their flight by moving from tree to tree.

Aviculture

The first recorded breeding of this species was by Mrs Johnstone at Bury St Edmunds, England, in 1901 when three young were reared. Remarkably the first success on the Continent was not until 1946 when Bruyneel of Belgium was successful.

As usual the early Australian breedings have gone unrecorded. The first recorded breeding was by R. J. Bellchambers of Humbug Scrub, South Australia in 1935, when two young were reared.

I consider the Major Mitchell to be the easiest to breed as well as the most commonly bred cockatoo in Australia. Aviculturists are continually successful with this highly desirable aviary bird throughout the Commonwealth.

My first success was in 1965 with a hand reared pair which were obtained from different sources. Both birds spent the early years of their lives, prior to coming into my possession, in "cockies' cages" as pets. They were housed in a small holding aviary, 1.8 m (6 ft) long, 90 cm (3 ft), wide and 1.5 m (5 ft) high. The nesting log provided which I now consider to be totally inadequate, was 30 cm (12 in.) internal diameter and 60 cm (2 ft) long, hung vertically with the entrance hole through the open top of the log. My sketchy notes of this breeding which somehow survived the last 22 years and my disastrous filing system, state that the first egg was laid on 26 September, the second egg on 29 September. One chick hatched on 23 October and the other egg was infertile. The chick grew very little in the first two weeks but the massive growth rate in the third week was remarkable. The chick was closed rung at three weeks of age. At four weeks old the chick was in advanced pin feather stage with its crest quite noticeable and my notes say that I was bitten when I handled it. When five weeks old the chick was almost fully feathered and at six weeks of age he was fully feathered and appeared ready to leave the nest. On 12 December, 1965, the baby left the log in perfect feather almost as brightly coloured as the parents and with dark brown to black eyes. This chick fledged in 50 days. The incubation period in this case, assuming that incubation started with the second egg, was 25 days.

This pair subsequently nested for several successive seasons. They were always single brooded and after this first nesting always flew two or three young per year. I eventually lost the hen from egg binding, which in this case I am sure was totally related to insufficient exercise due to inadequate housing. Other pairs in my collection housed in aviaries five to seven metres long have had no obesity or egg binding problems.

The lessons for us all to learn from my first breeding pair of Major Mitchells is that hand reared cockatoos can make good breeding birds. Compatible pairs will breed successfully in inadequate housing but in the long term the lack of exercise will be detrimental.

Adequate exercise is essential for a cockatoo's general health and well-being.

The following summary of breeding data from two pairs of the western subspecies of the Major Mitchell *Cacatua leadbeateri mollis* over three seasons, 1985, 1986, 1987, was gleaned from extensive information compiled by Neil Hamilton and Peter Brindly in Perth Zoo. It was kindly made available by John De Jose, Director.

Aviaries housing these pairs of cockatoos were 9 m (29 ft) long, 1.8 m (6 ft) and 2.2 m (7 ft 2 in.) wide and 2.3 m (7 ft 6 in.) high, with 2 m (6 ft 6 in.) of the rear end of the aviary roofed and sheltered.

Nesting logs are 60 cm (2 ft) long, with an internal diameter of 25 cm (10 in.). They have a perforated flat iron bottom and a flat iron top. Entrance is via a 15 cm (6 in.) "V" cut in the side of the log from the top. The logs are stood vertically on a 1.3 m (4 ft 3 in.) high metal stand situated on the right hand side of the open section and roughly in the centre of the length of the aviary. Protection from the weather is provided by a sheet of flat iron fixed to the top of the aviary.

Working of the log commenced during July and was followed by periods of copulation through the weeks preceeding egg laying. Eggs were laid during August and September with the incubation carried out by both the male and female. Clutches of two to four eggs were observed. Unfortunately it was not possible to record exact incubation periods.

Young were hatched during September and October and were always fed by both parents. Fledging was recorded in November and December, approximately six weeks after hatching. Clutches of one, two and three young were fledged during the three-year period.

One pair showed extreme aggression towards the keepers during the breeding season whereas the other pair were calm and only occasionally became a little aggressive.

Sexing

Major Mitchells are easily sexed by their eye colour, with females having a reddish light brown iris and the males a dark brown to black iris. Females also have a wider yellow band in the crest. The sex difference of

A Nominate subspecies *C. l. leadbeateri* (female) with yellow band in crest and *C. l. mollis* with all-red band in crest (male).

lings one and three days old.

C Nestlings three weeks old. Note how younger chick has slipped behind.

lings five weeks old. Younger chick catching

E Nestlings six-and-a-half weeks old just prior to fledging.

Plate 26: Major Mitchell Cockatoo — Subspecies and Nestlings.

the eye colour becomes apparent in immatures within the first twelve months, usually at around eight or nine months of age. Fledgling males are generally brighter than the females, particularly in the pink on the chest.

Display

Like most cockatoos the pair bond is strong. Established pairs are devoted and enjoy sitting close to each other with mutual preening being practised all year round. Display consists of the male approaching the female in an elated way with his body and crest erect while uttering his normal call in a highly excited manner, with his head bobbing and twisting almost simultaneously. The female usually responds by erecting her crest and bobbing her head.

Nests

Over the years my Major Mitchells have accepted both horizontal and vertical logs ranging from 30 cm (12 in.) to 45 cm (18 in.) internal diameter and 60 cm (2 ft) to 120 cm (4 ft) long. Generally I feel that vertical or near vertical logs are preferred. However, a particularly successful pair in my collection use a horizontal log which is 60 cm (2 ft) long and 30 cm (1 ft) internal diameter, with a natural spout entrance facing upwards at an angle of 45 degrees from the top of the log. Nests should be filled with up to 15 cm (6 in.) of fine wood shavings, peat moss or natural wood dirt.

Nesting and Hatching

I have found three eggs to be the average clutch size in my aviaries, although two eggs (particularly in dry seasons), and on occasions four eggs have been recorded. The eggs are usually laid at two or three-day intervals during the months of September and October. Should the first clutch be lost or the chicks taken for hand rearing at the early pin feather stage, a second clutch is sometimes laid in the month of November or as late as December.

Incubation is usually started with the second egg and is shared by the male and female. I have found that with most pairs the male incubates during most of the daylight hours and the female at night. My records show incubation periods of 24 to 26 days with an average of 25 days. The chicks hatch almost naked having just the slightest trace of a whitish whispy down. Early growth rate is usually slow, but between two to three weeks the chicks start to grow rapidly, their eyes open and early pin feather development is apparent. As stated previously young are fully feathered at six weeks and the records show fledging usually occurs between seven and eight weeks of age.

Fledglings, which are duller replicas of the adults with dark brown eyes, are often fed by their parents for some months after leaving the nest. However, a close watch must be kept on pairs with young flying, for any sign of aggression. If the parent birds do become aggressive the babies should be removed to a well protected holding cage and observed closely to ascertain if they are feeding themselves. If not, supplementary feeding with a syringe and crop needle may be necessary. Normally it is safe to remove young Major Mitchells from their parents four weeks after they leave the nest.

Much has been written, particularly in overseas avicultural literature, of the aggression shown by many male Major Mitchells towards their females, which often results in the female's murder. I have never experienced this behaviour and have always found that established pairs are particularly devoted to one another. Unlike many psittacines, I have not found it difficult to form compatible pairs and, in fact, almost without exception every pair placed together have accepted each other. The only murder recorded in my aviaries with this species was when two adult males who were housed together fought and one was killed. In view of the recorded acts of aggression, perhaps it is wise to keep a close watch on newly established pairs, particularly as the breeding season approaches.

Generally I have found this species to be a reliable breeder, that is, not subject to unpredictable behaviour such as leaving fertile eggs or deserting chicks, etc. Such occurrences are usually one-off things and may be initiated by some undetected trauma such as a night predator. On the other hand the problem may be related to an individual bad habit or vice in one or both of the pair concerned. Whatever the origin of the problem, we, as aviculturists, are often left to solve it in a hurry.

My good friend Jack Stunnell, who is a very experienced and versatile aviculturist, was confronted with one of these disasters a couple of seasons ago, when a pair of Major Mitchells left three fertile eggs. Jack wandered his aviaries with the cold eggs in his hand looking for a pair of birds that might continue the incubation. The only birds at the right stage of incubation were a pair of Indian Ringnecks, who accepted the eggs and duly hatched the chicks. They reared them until they advanced enough to be easily hand reared. Normally my comment on a breeding such as this would be that a lot of luck was involved but Jack is so constantly successful with unlikely avicultural feats, that I am sure that there must be more than luck involved in Jack's case

Major Mitchells normally won't breed successfully until three or four years old, although Harvey Oliver, now of Wagga, bred from a two-year-old aviary bred pair during the 1970's, whilst he was residing in the south-western suburbs of Sydney. The life span and breeding life of the Major Mitchell is similar to that of humans.

Mutations

I can find no record of a mutation having occurred in this species.

Hybrids

Hybrids have been recorded with the Sulphur-crested Cockatoo, the Short-billed Corella, Galah and the Great White Cockatoo.

SHORT-BILLED CORELLA

Plate 27: Short-billed Corella *Cacatua sanguinea*.

SHORT-BILLED CORELLA

(Cacatua sanguinea)

Derivation:

Cacatua — Latin from the Malay word *kakatua* for large parrot

sanguinea — Latin for blood red

Description:

Length 40 cm average
Weight 540 g average (males usually heavier than females)
Appearance as in Plate No. 27

Classification

I am not prepared to accept the current classification of this species, as adopted by Forshaw in his second edition of "Australian Parrots", which not only continues to deal with the Western Long-billed Corella and the Short-billed Corella as members of the same species, but lists the Western Long-billed Corella *(Cacatua pastinator)* as the nominate race and the various Short-billed Corellas as subspecies. Hence I am adopting the previous scientific name *sanguinea*, originally bestowed on it by Gould. These classification changes eliminate the confusing and unsatisfactory situation of the Short-billed Corella being a subspecies of the Western Long-billed Corella.

There are three recognised subspecies of the Short-billed Corella. (See Plate No. 28.)

Cacatua sanguinea sanguinea is the largest and whitest of the subspecies, having only a little pale orange-red around the beak and a paler blue periophthalmic eye ring.

Cacatua sanguinea gymnopis is slightly smaller than *C. s. sanguinea*, a little more off-white in colour, with more extensive and pronounced orange-red around the beak area and a slightly darker blue periophthalmic eye ring.

Cacatua sanguinea normantoni is the marginally smallest subspecies similar in colouration to *C. s. sanguinea* with a medium blue eye ring.

Earliest Report

This species appears to be the first Australian parrot sighted and recorded by a European. In August 1699, William Dampier observed Short-billed Corellas on the island group known as the Dampier Archipelago and described the sighting in his "Voyage to New Holland" 1703.

Gould described this species from specimens taken at Port Essington (which would have been *C. s. sanguinea*) in the "Proceedings of the Zoological Society", London 1842, page 138, February 1843, as *Cacatua sanguinea*.

The subspecies *C. s. gymnopis* was described by Sclater in 1870 from a living specimen and from skins in the British Museum taken by Captain Charles Sturt in South Australia. This cockatoo was apparently first illustrated in John Goulds' "Birds of Australia" 1848, Volume 5, Plate 3.

Range, Habitat and Field Notes

The Short-billed Corella ranges over much of the Australian continent, but is absent from Tasmania. It is a particularly successful species which is constantly extending its range, with the assistance of the establishment of permanent water for stock grazing purposes, in semi-arid areas where watering places were previously unreliable.

This cockatoo can usually be observed in good numbers and often in huge flocks throughout its range.

The exact range of the three subspecies is still largely undetermined and likely to remain that way unless intensive field work is carried out, as identification of the subspecies of this cockatoo by normal field observation is extremely difficult.

C. s. sanguinea, the nominate subspecies, is possibly an isolated population, ranging across the extreme north-western region of Australia, from the western side of the Gulf of Carpentaria, westward to the Kimberlies and the locality of Broome, Western Australia. Its southern progress appears to be restricted by the arid interior of the Northern Territory in the east and by the Great Sandy Desert in the west. Most offshore islands in this north-western region also support this subspecies.

C. s. gymnopis is by far the most widespread subspecies, with distinct eastern and western populations, both of which are extending their range southward. The eastern population extends through the drier inland areas of south-eastern Northern Territory, central and south-western Queensland, western New South Wales, north-western Victoria and the eastern half of South Australia. The western population extends south and west from the Great Sandy Desert, and west from the arid inland areas of Western Australia, such as the Gibson Desert and the Great Victoria Desert, to the coast and many offshore islands, then south to the Moora and Lake Cowcowing area.

C. s. normantoni's range is uncertain, but it is known to occur in southern New Guinea, the western Cape York region and the south-eastern areas of the Gulf of Carpentaria.

Isolated breeding populations, which are probably the result of escaped or released cage birds, occur in the vicinity of Perth, Western Australia; Brisbane, Queensland; Sydney and Gosford, New South Wales; and Melbourne, Victoria.

I have observed *C. s. sanguinea* a number of times in the northern regions of the Northern Territory, but the sighting that remains most vivid in my mind, probably because it is a more recent observation (July 1986) occurred at Foggs Dam, a renowned bird watching area south-east of Darwin. We arrived at Foggs Dam in the late afternoon of a hot sunny day (by our southern standards), with the intention of observing water birds which are in vast numbers in the area. The dam is a large, crocodile infested body of water divided in half by a roadway which provides easy access and simplifies birdwatching. As we approached, the constant raucus calling heralded the presence of a flock of several hundred Short-billed Corellas. A hundred or so were busily digging for some sort of marine or insect life, or possibly vegetable matter, on mud flats at the near end of the dam, which had been left exposed by the receeding water level as the dry season progressed. The cockatoos, many of whom where soiled around the face and chest, were constantly coming and going to and from the mud flats. As we progressed along the mile of roadway bisecting this dam, with our binoculars glued to our eyes, revelling in the large and varied number of species to be seen, we found large groups of Corellas perching in the low Eucalypts which were lining the side of the roadway.

Most of the flocks were of adult or sub-adult birds but here and there an occasional youngster could be heard before being seen as it cried and begged incessantly for food. Every now and then a flock of 50 or so would noisily vacate their tree and fly to another more to their liking, often almost blotting out the green foliage of the small trees with their white plumage.

Another memorable sighting of this subspecies was of a small flock at Port Essington, which is almost as remote and inaccesable now as it was 150 years ago, when the first specimens were collected there for John Gould.

My recollections of *C. s. gymnopis* are mainly of large noisy flocks, often around water, in hot, dry places. Yet the experience which will always remain most vivid in my mind occurred almost twenty years ago on my first visit to the Northern Territory. My friend Ron Pascoe, myself and our families were driving our vehicles on a track which crossed the dry and dusty, flat open plains that extend from Anthony's Lagoon to the Stuart Highway, when we came across a very large windmill. This mill would pump water into a "Turkey's nest" tank (an elevated earthern water storage receptacle) when the wind blew and the tank in turn supplied water to the stock drinking troughs. Due to the dry conditions at the time, many thousands of Short-billed Corellas had gathered in the area to be close to the permanent water supply.

The day was hot and still without a breath of wind, hence the windmill was not operating, and perched on every available inch of the mill were thousands of Corellas. Every horizontal and near horizontal section of the structure, including the wind vanes, had Corellas stacked shoulder to shoulder, all striving to retain their position as numerous others tried to squeeze in or dislodge another already perched. This normally stark metal windmill had become a writhing, squabbling, screeching, white mass of Short-billed Corellas.

My most lasting impression of *C. s normantoni* is again one of heat, dust and Corellas, as well as Fork-tailed Kites and the Gulf of Carpentaria. It is now over 25 years since I had my first view of the Gulf of Carpentaria and on the way my first visit to the then harsh, dry and dusty town of Normanton, situated on the Norman River some forty or fifty kilometres inland from the south-eastern corner of the Gulf. At the time this town supported a large flock of Short-billed Corellas *(normantoni)*, large numbers of Fork-tailed Kites which scavenged the streets like sparrows do in the south, and numerous Aborigines. The Corellas also scavenged in the town, as well as feeding on the surrounding country.

More recent visits to Normanton have revealed a more pleasant and hospitable town, which can still boast a generous population of Corellas, Kites and Aborigines. I assume that this subspecies name is derived from its occurrence in the Normanton region.

Short-billed Corellas occur in a variety of habitats, including open forest, scrubland, open and tree dotted grassland, tree lined water courses, mangrove swamps, arid semi-desert country and so on, but their existence in a given area is dependant on a permanent water supply. This species is common around inland towns and their local garbage tips, where they forage for food scraps.

Their diet is as varied as their habitat and consists of seed, fruits, berries, nuts, blossoms, roots and bulbs of native and introduced grasses, ground plants, shrubs and trees as well as insects and insect larvae, etc. Crop damage is often inflicted by this cockatoo, particularly in grain growing areas, such as the rice fields of Northern Territory.

I have often observed Short-billed Corellas feeding in the company of Galahs and occasionally with Sulphur-crested Cockatoos.

Information derived from tagged birds suggests seasonal movements of 200 kilometres or more after the breeding season, to established feeding areas, then later returning to their original breeding grounds. Birds studied in other areas moved only locally after breeding.

The flight of this cockatoo consists of varying periods of deep wingbeats followed by short periods of gliding flight.

Breeding in the Wild

The preferred nesting site is usually a hollow limb or hole in the trunk of a Eucalypt tree at recorded heights of 3 m (10 ft) to 10 m (38 ft) above the ground, with depths of hollows of 50 cm (20 in.) to 1.2 m (4 ft) and internal diameters of about 35 cm (14 in.).

Nesting may occur as early as February or March in the far north, to July and August in the central latitudes and as late as October in the south. Recently fledged babies I observed near Darwin during July 1986, would have to be the product of eggs laid during March. Whereas in years gone by, well feathered nestlings taken for the pet trade in western New South Wales and South Australia, were always available during November and early December, which would suggest nesting commenced during the later half of August or September.

Both male and female prepare the nest hollow by chewing around the entrance and interior of the nesting chamber. Two or three white, oval eggs are laid on a rotten wood dirt and wood chip base, and are incubated by the male and female, with the male carrying out the majority of the daytime duties while the female incubates during the night.

The chicks hatch with a pale lemon-yellow down and are fed by both parents in the nest as well as for some weeks after fledging. Recently fledged young Corellas can be detected a considerable distance away by their constant wheezing and whining as they beg for food.

After the breeding cycle is completed, the breeding pairs and their young usually join the non-breeding flocks to forage for food, either locally or at habitual feeding grounds.

Aviculture

Prestwich, in his "Records of Parrots Bred in Captivity", states that the first breeding success occurred in London Zoo when one youngster was reared in 1907. He also records many other successes such as: Rome Zoo, 1923, three young reared; Holland, 1926, Blaauw reared one; San Diego Zoo reared their first young in 1929 and reared approximately 50 during the following 22 years, although the bulk of these were hand reared.

The first recorded Australian breeding was Adelaide Zoo in the 1949 season, when two pairs nested. One pair, under segregated conditions, hatched but failed to rear their young, while the other pair, who were housed in a large flight aviary with 30 to 40 other mixed cockatoos, were successful. One young was reared and left the nest in mid-January 1950.

My first success occurred in 1973 with a pair I purchased from Alan Parmenter, a prominent Sydney bird dealer, a few months prior to their breeding. I can remember going to Alan's premises, then in Hatfield Road, Canley Heights, and seeing this immaculate and obviously devoted pair of Short-billed Corellas sitting in a small cage. Although I needed a pair of Corellas like I needed another hole in my head, I could not resist such a delightful pair of birds and took them home with me.

The only space I had available for them was a small holding aviary 1.8 m (6 ft) long, 1.8 m (6 ft) high and 90 cm (3 ft) wide. The cage was fully roofed and enclosed on the back as well as the back half of both sides. A hollow log which was 60 cm (2 ft) long, with an internal diameter of 30 cm (1 ft), the entrance via the top and filled to a depth of 10 cm (4 in.) with wood shavings, was hung at the rear of the shelter.

As the spring approached, both birds were seen to go in and out of the log, then, in mid-September, two eggs were observed and incubated by both male and female. During the second week of October two chicks were hatched, after an estimated incubation period of 26 days.

Approximately seven weeks after hatching both youngsters fledged in beautiful condition. The parents continued some feeding of the fledglings for eight weeks, then the young were deemed independent and removed.

This pair were fed a basic dry diet of sunflower seed together with a daily ration of silverbeet, green peas, apple, brown bread and milk arrowroot biscuit, all of which were increased considerably when young were being reared.

Two young were normally fledged each season except for one occasion when three left the nest. After breeding from this pair for five consecutive seasons in their small holding aviary, I gave the pair to Peter Chapman, who, I believe bred from them before passing them on to yet another cockatoo breeder. They were a pair in a million — the pair that we have all dreamt about, that never did a thing wrong and always reared successfully with a minimum of fuss.

During 1985, 1986 and 1987 seasons Jim Gill reared young Short-billed Corellas from a pair housed in a suspended wire aviary, measuring 3 m (10 ft) long, 1.2 m (4 ft) wide and 1.2 m (4 ft) high. A hollow log 75 cm (2 ft 6 in.) long, 37 cm (15 in.) internal diameter, with an entrance hole in the side at the top of the log, filled to a depth of 15 cm (6 in.) with wood shavings, hung on the outside of the aviary in a vertical position, was provided.

Both birds worked the log and laid their first egg on 17 August, 1986, followed by a second, third and fourth egg on 19, 21 and 24 August respectively. Incubation was carried out by the male and female with the male incubating during most of the day. The first two eggs hatched on 14 September, 1986, after an incubation period of 26 days, followed by the third egg on 17 September after 27 day's incubation; the fourth egg failed to hatch. The chicks hatched with yellow down, horn coloured beaks and black eye spots. The eyes started to open and early pin feathers could be seen at two weeks old, then at three weeks the crest feathers started to appear. When four weeks old, pin feathers were well developed all over the young and by six weeks they appeared almost fully feathered. Fledging occurred on 18 and 20 November after 65 and 64 days in the nest respectively. Jim fed a basic diet of sunflower seed together with the only supplementary foods this pair will accept — corn on the cob and seeding grasses. The refusal to eat a wide range of nutritious supplementary foods probably reflects thc long fledging period of the young reared by this pair.

A Nominate subspecies *C. s. sanguinea*. Note little red and paler blue eye ring.

B *C. s. gymnopis*. Note more red and dark b eye ring.

C *C. s. normantoni*. Note little red and medium blue eye ring.

D Western Long-billed Corella *C. p. pastinator*.

E Eastern Long-billed Corella *C. tenuirostris*

Plate 28: Comparisons of Short-billed Corella subspecies and Long-billed Corellas.

Sexing

Visual sexing of this species is virtually impossible. Although my breeding pair were an obvious pair while together, when separated I could not tell one from the other. The male usually has a slightly larger head and upper mandible, while the female is generally, marginally smaller, although these differences are variable. Surgical sexing is essential if pairs are required for breeding. Immature birds are even more difficult than adults to visually sex.

Display

Display commences with the male screeching excitedly, while spreading his wings, erecting his crest, then approaching the female with his head bobbing, tail fanned and occasional bowing, while uttering an unusual, almost yodelling call.

Nests

The most preferred nest is a vertical hollow log about 75 cm (2 ft 6 in.) long, 30 cm (12 in.) to 35 cm (14 in.) internal diameter, filled with 15 cm (6 in.) of wood dirt or wood shavings, with a natural spout or hole, at or near the top of the log. I prefer all nesting logs hung under shelter for protection from the elements. This cockatoo is normally not difficult to satisfy with nesting facilities. Some breeders supply taller logs stood on the ground and filled to within 60 cm (2 ft) of the entrance hole.

Nesting and Hatching

Breeding intentions are suggested by noisy and excited behaviour and increasing attention to the nesting log with both male and female chewing and fussing around the entrance as well as inside the hollow. At times excessive excavation of the filling may necessitate more being added.

In the Sydney area this lead up to breeding usually occurs in August or September with eggs being laid in September or October. Of course, birds breeding in aviaries further north normally breed earlier and in the south, later.

Two or three and occasionally four white oval eggs laid at two or three-day intervals form the usual clutch. Incubation is commenced by the female, usually after the laying of the first or second egg, then after the first day or two the male assists with incubation during the daytime. The average recorded incubation period is 26 days.

A Newly hatched chicks.

B Nestlings three weeks old.

C Nestlings five weeks old.

D Nestlings six weeks old, just prior to fledgi

Plate 29: Short-billed Corella — Nestlings.

Chicks hatch with yellow down, horn coloured beaks and black eye spots. The growth rate is slow at first, the eyes start to open at about 14 days old and early pin feather development can be seen soon after. When four weeks old the chicks are a mass of spikey pin feathers and are fully feathered at six weeks.

Fledging periods are variable but usually occur between six and eight weeks old.

The young are fed in the log by the female and the male, who continue to feed their young for several weeks after they leave the nest. A close vigilance must be kept on the behavioural attitudes of the parents towards their young. The moment any sign of aggression is detected the young must be removed, even if supplementary feeding is required, for these early signs of aggression can quickly lead to maiming or even murder.

Mutations and Hybrids

I have been unable to find records of mutations occurring.

Hybrids between this species and Sulphur-crested Cockatoo, Major Mitchell Cockatoo, Galah and Gang Gang have been recorded.

WESTERN LONG-BILLED CORELLA

Plate 30: Western Long-billed Corella *Cacatua pastinator*.

WESTERN LONG-BILLED CORELLA

(Cacatua pastinator)

Derivation:

Cacatua — Latin from the Malay word *kakatua* for large parrot

pastinator — Latin for one who digs and trenches a vineyard

Description:

Length 42 cm average
Weight 700 g average (males usually heavier than females)
Appearance as in Plate No. 30

Classification

Formerly this Corella was classified as a subspecies of the Eastern Long-billed Corella, *Cacatua tenuirostris*, with the scientific name of *Cacatua tenuirostris pastinator*. This procedure was followed by Forshaw in his first edition of "Australian Parrots."

Since then, following work carried out by Schodde on the Corella group, it has become popular to classify the Western Long-billed Corella as one of the subspecies of the Short-billed Corella. In the second edition of "Parrots of Australia" Forshaw classifies the Western Long-billed Corella as the nominate subspecies of the Short-billed Corella using the scientific name *Cacatua pastinator pastinator*. In other words the Short-billed Corella has become a subspecies of the Western Long-billed Corella, a situation that must be considered as unsatisfactory.

Obviously I am only a lay birdman — an aviculturist whose only qualifications are experience and a big mouth, but I fail to see how the Western Long-billed Corella and the Short-billed Corella, who have an overlapping range, have managed to remain distinct for thousands of years if they are NOT separate species.

Aviary studies have shown distinct differences in behavioural patterns, temperament and calls between the Western Long-billed Corella and the Short-billed Corella as well as a shorter incubation period for the Western Long-billed Corella (23-24 days) than any other Australian member of the Cacatua genus. Surely if these Cockatoos were in fact members of the same species they would retain the same incubation period.

Forshaw points out that specimens from the southern parts of the range are larger and have longer bills than the northern birds and suggests that this is due to the gradual intergration to the Short-billed Corella in the north. I would suggest that perhaps the decrease in the length of the bill could be related to habitat and hence feeding variations from south to north. This has already been observed in two other species of cockatoos endemic to the region, namely the White-tailed Black Cockatoo, *C. baudinii*, and the Red-tailed Black Cockatoo, *C. m. naso.*

It is interesting to note that the people who are so intent on the abolition of the species status of the Western Long-billed Corella are now suggesting the elevation of the Long-billed White-tailed Black Cockatoo, *C. b. baudinii*, to full species status, which is the exact opposite of their treatment of the Western Long-billed Corella.

Fashions in bird taxonomy appear to change just as often as they do in ladies' fashion and possibly for just as obscure reasons. The trend in the early part of this century was to create as many new species as possible, often for no better reason than some minute racial difference. Now the wheel has turned and the trend is to lump all subspecies and even species together. For example, there is a current move to reduce the eight recognised species of the Rosella family to three species — the white cheeked group, the blue cheeked group and the yellow cheeked group.

I have always been amazed how the ornithological community accepts without question the often unwarranted classification changes inflicted on them by taxonomic specialists within their midst. The avicultural community does not accept these changes so readily, obviously because the close association they have with their birds allows them to recognise subspecific differences and appreciate them.

During my own limited field observations of this cockatoo I saw it living in the same locality as the Short-billed Corella each type of Corella apparently filling its own ecological niche, with no association between them.

In summary, I consider the evidence presented sufficient to separate the Western Long-billed Corella from the Short-billed Corella. As it is generally accepted as not being closely related to the Eastern Long-billed Corella, the Western Long-billed Corella must be given separate species status under the previous scientific name of *Cacatua pastinator*.

Earliest Report

This cockatoo was described by Gould in the "Proceedings of the Zoological Society", London 1840, page 175. He gave it the Scientific name *Licmetis tenuirostris pastinator*, as a subspecies of the Eastern Long-billed Corella, *Licmetis tenuirostris*. Later Gould described this cockatoo as a distinct species in his "Handbook of the Birds Of Australia" 1865, Volume 2, page 12 as *Licmetis pastinator*, based mainly on his observations of living specimens of both Eastern and Western Long-billed Corellas housed in the menagerie of the Zoological Society of London.

The first illustration of this cockatoo appears to be a plate by F. W. Frohawk featured in Gregory Matthews' "Supplement to the Birds of Norfolk and Lord Howe Islands" 1936 which has added some supplementary birds of New Zealand and Australia. It is labelled as the Long-billed Cockatoo, *Licmetis pastinator*, Plate 62, page 24.

Range, Habitat and Field Notes

The Western Long-billed Corella's range is apparently restricted to two populations in the south-western corner of Western Australia. Information compiled by the RAOU for "The Atlas of Australian Birds" indicates that these populations are separated at the approximate latitude of 32°S, roughly in the Mt Cooke, Pingelly, Corrigin regions. This current range appears to extend from the Pemberton-Lake Muir region in the south to Geraldton on the coast in the north, then inland to the vicinity of Lake Monger and as far east as the Lake Deborah district. This is only a remnant of a once widespread range over the coastal plains and woodlands.

My first and only sightings of this cockatoo in the field occurred over 20 years ago. I can still remember the thrill of seeing the first pair feeding on grassland, industriously employed digging for some root or bulb.

At this stage the Western Long-bill's numbers were possibly at an all-time low hence I did not find it to be one of the easiest cockatoos to locate. Fortunately, since that time the populations have gradually increased and current information suggests that numbers have improved considerably in both groups, obviously aided by legal protection.

The preferred habitat is open forest country, timbered water courses and farmland where adequate belts of large trees have been retained. Of course all suitable habitat is dependent on permanent water supply.

Foods consist of seeds, roots and bulbs of various native and introduced grasses and ground plants as well as herbage and insects. The bulb of the introduced onion grass is a particular favourite of this cockatoo.

After breeding, flocks of young and adult birds are formed and commence nomadic foraging for food. These localised movements are unpredictable and related to the availability of suitable food sources. Grain crop damage sometimes occurs during these forays and unfortunately retaliatory action is often taken by the farmers. Adult birds usually return to their established localities during autumn or early winter.

The huge flocks of the time of the first white settlement are gone, although the current tendency of the Western Long-bill to adapt to feeding on grazing and farming land, together with the availability of permanent water, does present fresh hope for this species. Let us hope that a revival of the former numbers of this cockatoo in farming districts does not provoke similar response to that imposed on the Eastern Long-bill by the Victorian wildlife authorities.

The flight of this cockatoo is similar to the Sulphur-crested Cockatoo, but with a greater depth of wing beat followed by short periods of gliding flight.

This cockatoo is identified in the field from the Short-billed Corella by its larger size, long, slender upper and lower mandible, slightly larger crest and more extensive and pronounced brick red colouring between the eyes and around the upper mandible.

Breeding in the Wild

The nesting site is usually a large hollow limb or hole in the trunk of a Eucalypt tree with heights of 3 m (10 ft) to 9 m (30 feet) and depths of hollows of 60 cm (2 ft) to 1.2 m (4 ft) having been recorded.

Breeding usually occurs from August to October when two to four (usually three) white oval eggs are laid in the hollow which is prepared by the male and female by chewing the wood particles and turning over the wood dirt in the base of the log.

Incubation is carried out by both the male and female with the male sitting during the majority of the daylight hours and the female incubating for the balance of the day and all night. The chicks hatch with a sparse pale lemon down and are fed in the nest by both parents. The parents continue to feed their crying, begging young for some weeks after fledging.

When the young commence to feed themselves, family parties and small flocks are formed as they wander nomadically in search of good feeding grounds.

Aviculture

Although this cockatoo is not the ugliest of our Australian cockatoos it must rate second to the Eastern Long-billed Corella.

I can find no early avicultural reference to this species — it was apparently unknown to early English and European aviculturists and as far as I can determine still is today.

Recent Australian avicultural literature mentions the Western Long-billed Corella but provides no breeding records. Forshaw states that this cockatoo has been bred in zoos and in a few private aviaries, but provides no details.

The first recorded breeding of this cockatoo occurred in Perth Zoo during the 1971-72 season and was listed in the Annual Report, 30 June, 1972. Unfortunately, few details were documented but apparently two chicks were hatched; one died whilst small and the other was taken for hand rearing when almost fully feathered because the parents were losing interest in it. The youngster developed into a beautiful bird and eventually proved to be a male.

Two young Western Long-bills were bred, apparently parent reared in the 1973-74 season and recorded in the zoo's Annual Report, 30 June, 1974. Yet another was bred in 1976-77 season and recorded in the June 1977 Annual Report.

The first well documented breeding of the Western Long-bill appears to be that of Jim Gill in 1985. The pair were housed in a suspended aviary which was 3 m (10 ft) long, 1.2 m (4 ft) wide and 1.2 m (4 ft) high and situated in an enclosed sheltered area with only about 1.2 m (4 ft) exposed to direct weather conditions. The nest consisted of a large hollow log 1.2 m (4 ft) deep and with an internal diameter of 50 cm (20 in.). There is a natural entrance at the top of the log. The log is suspended on the outside of the aviary and filled to a depth of 10 cm (4 in.) with wood shavings.

The pair were hand reared birds and were supposedly only two years of age. An egg was laid on 1 September, 1985, which was destroyed by the male within a few hours. A second egg was laid on 3 September and was removed and fostered under a pair of Short-billed Corellas. A third egg was laid on 7 September and was also fostered under the Short-billed Corellas. Fourth and fifth eggs were also laid at four-day intervals but were not rescued before the male consumed them. Both eggs hatched after an incubation period of 24 days under the Short-billed Corellas. The chicks were covered in a sparse pale lemon down. The second chick died after 40 hours. The other chick was not growing as well as its Short-billed Corella nest mates. At 10 days its eyes were partially opened and it was removed for hand rearing. Pin feathers started to appear at 18 days of age. It continued to grow well until it was finally weaned just prior to Christmas some 11 weeks after hatching.

In the spring of 1987 Jim Gill presented me with a fresh Western Long-bill egg which was removed from the above pair of birds. After weighing the egg (18.89 g) I placed it in a Marsh RollX incubator which already contained three exotic parrot eggs. The temperature setting was 98.5°F with a wet bulb reading of 90°F which I consider to be too humid for Australian cockatoo eggs and more suited to the exotic parrot eggs being incubated at that time.

The egg was turned end for end twice daily, besides the normal turning carried out by the automatic turner. The weight of the egg was checked every few days and was found to be losing too little weight due to the excessive humidity. After 22 days incubation the egg weighed 17.3 g which I calculated was 1.41 g too heavy.

Chipping of the egg was observed on 5 October, 1987, after 23 days incubation, but 24 hours later the chick had made very little progress so I decided to assist the chick. Over an eight-hour period the top of the shell was picked away little by little, with a small pointed knife, until the head, wings and top half of the chick was exposed and clear of the rest of the shell. It soon kicked its way clear.

The excessive humidity obviously toughened the skin-like membrane which lines the shell thus preventing the chick from hatching. A more suitable wet bulb thermometer reading for incubating cockatoo eggs is 86°F.

Two hours after hatching the chick weighed 13.95 g and was covered with a fine pale yellow down and had a noticeably narrow elongated upper mandible and black eye spots. (See Plate No. 31A.) The chick was fed a small quantity of crop wash (see Management chapter) from a small shute-like spoon until the crop appeared slightly inflated. The crop wash was taken from the crop of a healthy adult male Galah and stored in a refrigerator between feeds. I have found the feeding of a crop wash to newly hatched chicks to be the most reliable method of providing the essential bacteria necessary for successful food digestion.

Feeding then continued every two hours with equal parts of strained baby apple and strained baby mixed vegetables together with a few drops of crop wash, all mixed together with previously boiled water. It was fed as a runny consistency, just warm to touch, and enough so as to slightly extend the crop.

When after four or five feeds the crop was still emptying completely every two hours an equal part of rice cereal was introduced into the mixture and continued for the next 24 hours. Then as digestion was still being completed between feeds the rice cereal was replaced with Farex which has a higher protein content. This mixture was continued until the chick was three days old when the mixed vegetable was replaced with an equal part of finely ground egg and biscuit canary rearing food and 20% protein chicken starter crumbles in equal proportions.

For the first three days feeding was continued every two hours from 6 a.m. until 10 p.m. plus one feed at 2 a.m. during the night. After the first three days the 2 a.m. feed was discontinued. As the chick's crop capacity increased the time between feeds was extended to two and a half hours, then three hours and so on.

When one week old the diet was changed to equal parts Farex, Canary egg and biscuit and 20% chicken starter crumbles together with half a part baby apple and half a part sunflower meal. Then gradually over the next week the baby apple was discontinued and the sunflower meal was increased to one equal part.

From three days old a calcium supplement was added to the first feed of the day and a multi-vitamin and mineral supplement added to the last feed of the day.

This diet was continued until independence with the number of feeds per day being gradually reduced from five to four to three then two and finally one feed per day, given last thing at night as the chick was being weaned on to dry seed, sprouted seed, green peas, apple, brown bread and silverbeet.

The early weeks of development of this hand reared chick were slow. As mentioned the weight at hatching was 13.95 g, 30 hours later it weighed 17.5 g, at three days of age only 19.5 g, at five days 24.3 g, then at 10 days 36.65 g. From then on the daily gains increased regularly until a month old when suddenly the weight gains increased dramatically and the end result was a superb youngster. Throughout the entire rearing period the chick was active and healthy and apparently unaffected by the initial poor weight gains.

A few hours after it hatched the chick was placed in a brooder adapted from a Roll X incubator. It was kept at a constant temperature of 97°F which was recorded inside the aerated plastic margarine container where the chick was held to protect it from the forced draught within the brooder. After three days the temperature was reduced by 1°F per day until 90°F was reached when the chick was 10 days old. When the chick was at early pin feather stage about three weeks of age, it was removed from the brooder into a cardboard box with a vertical heating pad, as described in the Management chapter.

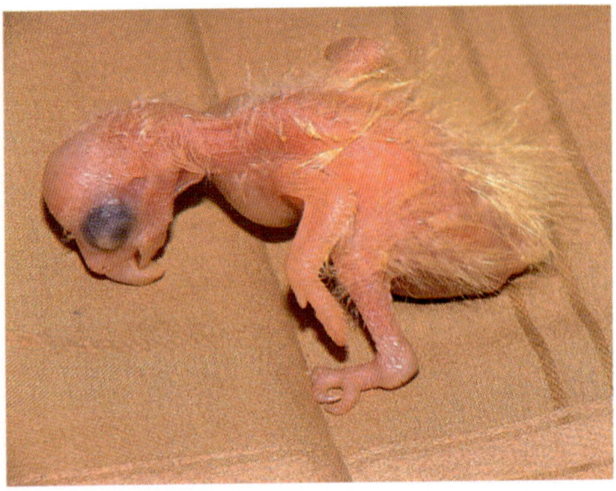

A Newly hatched chick. Note length of upper mandible.

B Nestling three weeks old.

C Nestling four weeks old.

D Nestling five weeks old.

E Nestling six weeks old, just prior to fledging.

Plate 31: Western Long-billed Corella — Nestlings.

At eight months of age this youngster weighed 591 g and was found to be a female when surgically sexed. Two similarly aged hand reared juveniles imported from Western Australia weighed 520 g and 600 g and both proved to be males.

Sexing

Visual sexing of this cockatoo is difficult. Males are often larger with a more robust head and a longer, thicker upper mandible as well as a more extensive, brighter brick red area between the eyes. These differences can only be treated as an indicator and not the rule.

Surgical sexing can save many wasted breeding seasons and must be considered the only sure method for this species. Immatures are even more difficult to visually sex.

Display

The male struts towards the female with his crest erected, while excitedly uttering his almost yodelling-like call whilst fanning his tail and bobbing his head.

Nests

The preferred nest appears to be a vertical or near vertical hollow log, about 90 cm long (3 ft), 30 cm to 38 cm (12 in. to 15 in.) internal diameter, and filled with up to 30 cm (12 in.) of wood dirt or wood shavings. A natural entrance spout or hole on the side of the log near the top is preferred. All nesting logs are best hung under shelter for protection from rain and heat.

This cockatoo does not appear to be as difficult to satisfy with a nesting site as the Eastern Long-bill.

Nesting and Hatching

The intention to breed is usually heralded by noisy courtship displays with increased and excited interest in the preferred nesting hollow. Both the male and the female are continually in and out of the log, chewing a little at its sides and excavating some of the filling. Should too much filling be removed more must be added to prevent damage to the eggs.

In the Sydney region of New South Wales breeding usually occurs during springtime. The few breeding records available indicate that egg laying normally commences in the first week of September. Three to five white oval eggs form the clutch and are usually laid at two to four-day intervals. We have recorded fresh egg weights of 15.3, 17.66, 18.89, 19.58 and 22.09 g. Incubation sometimes commences with the laying of the first egg but usually starts with the laying of the second egg. Incubation periods of 23 and 24 days have been recorded. The chicks hatch with a pale yellow down, a noticeable long, slender horn coloured bill and black eye spots.

The early growth rate is slow, the eyes commence to open between 10 and 14 days and pin feathers start to appear at about 18 days. The first sign of crest feathers are noticeable at 20 days.

When four weeks old, feather development is well advanced and chicks are fully feathered at six weeks of age, with fledging occurring between six and a half weeks and eight weeks of age. Immatures are similar to adults with less red colouring between the eyes and above the beak area and with a paler blue eye ring.

The young are fed by both parents in the nest and feeding is continued for at least four weeks after fledging, with the male often becoming more attentive than the female. Most young are independent a month after leaving the nest and can be safely removed from the parents 10 weeks after fledging.

Should any sign of aggression be observed from either parent towards the young it is advisable to remove the youngsters immediately, even if supplementary feeding is required.

Mutations and Hybrids

There are no records of either mutations or hybrids having occurred with this species, although I am sure hybrids must have been bred in Western Australia, but have passed without documentation.

EASTERN LONG-BILLED CORELLA

Plate 32: Eastern Long-billed Corella *Cacatua tenuirostris*.

EASTERN LONG-BILLED CORELLA

(Cacatua tenuirostris)

Derivation:

Cacatua — Latin from the Malay word *kakatua* for large parrot

tenuirostris:

tennis — Latin for slender

rostrum — Latin for bill

Description:

Length 39 cm average
Weight 650 g (males usually heavier than females)
Appearance as in Plate No. 32

Classification

The Eastern Long-billed Corella has been ranked as a species of the genus Cacatua by most recent authors, so rather than create confusion I have submitted to the authorative opinions relating to this species. I personally believe that if the Galah warrants the reintroduction of the monotypic genus Eolophus, then surely the Corellas should be separated from the other White Cockatoos, as explained in the introduction to the Cacatua.

This species is a specialised Corella which is unlike all other Corellas, even the apparent similarity of the Western Long-billed Corella is related to parallel evolution, rather than a close relationship to the eastern species. The distinctive long, slender upper mandible of both these cockatoos has developed to assist the digging of roots and bulbs in the almost totally terrestial feeding mode.

There has been no subspecies recorded in the limited and shrinking range of this cockatoo.

Earliest Report

This species of cockatoo was described by Kuhl in Nov. Act. Phys. Acad. Leop. Carol, Vol. 10, page 88, 1820, from specimens in the Paris Museum and at Brookes in London. It was given the scientific name of *Psittacus tenuirostris*. The first illustration of this species appears to have been by Temminck who described and portrayed it in 1819 as the Nasecus Cockatoo, *Psittacus nasica* (Dr Carl Russ, The Speaking Parrot, 1884).

A specimen of this species in the London Museum is said to have been taken at Port Phillip Bay (Victoria) on 27 April, 1802. This is well outside this cockatoo's present range. No doubt the Long-billed Corella previously enjoyed a far greater range than it does at present, although obviously many of the widespread early recorded sightings of this bird relate to the Short-billed Corella.

Range, Habitat and Field Notes

The Eastern Long-billed Corella has a restricted range in south-eastern Australia encompassing south-western New South Wales, western Victoria and south-eastern South Australia. It extends from the Coorong region of South Australia in the west, to the vicinity of Wagga Wagga, New South Wales, in the east, then south to the western Victorian coastline, almost to Port Phillip Bay and north to the Murray River system around Mildura.

During the last century this cockatoo's range extended along the Darling and Murrumbidgee River Systems and to the eastern side of Port Phillip Bay. This indicates that a massive decline in numbers and reduction of the area inhabited has occurred in this species. At the same time most other cockatoo species of the open forests and plains were and are extending their range and increasing in numbers.

A small isolated population has been observed around McLaren Flat in South Australia which is believed to have originated from aviary escapees. There are also established populations in the vicinities of Perth in Western Australia and Sydney in New South Wales, which have obviously originated from aviary or more precisely pet cage bird escapees.

Yet another larger established breeding population exists in the Gosford-Wyong district north of Sydney in New South Wales. My good friend Robert Adams of Empire Bay in this district recently observed a pair of Eastern Long-bills, accompanied by two newly fledged youngsters feeding at a bird feeding station on his property. The parents were still feeding these offspring which left no doubt in Robert's mind that they were recently fledged wild young.

Undoubtedly much of the blame for the establishment of these recent, unwanted and damaging populations far from the original distribution of this species, can be attributed to the irresponsible practices of some of our state wildlife authorities. By allowing the legalised trapping of supposed pest species for the aviary and pet trade without proper studies of the population dynamics could well contribute to this bird's extinction, as well as creating environmental damage in their new locations. This trapping is no more than a misguided attempt to control odd localised population explosions. The procedure works something like this: Trapping permits are issued to licensed trappers by the wildlife authorities to take a specified number of the offending species (which in the case of this cockatoo was vast numbers) from the wild to be sold into the pet trade. Because these trapped birds are usually common aviary or cage birds in their state of origin, they are usually exported legally to other states. These wild birds will usually adapt to aviary life but are totally unsuited for pet cage birds. Most new owners persevere with their wild birds for a few weeks or months then give them up in disgust and they are pushed aside for a new toy. Often these birds are released by their owners as an easy means of ridding themselves of these ungrateful and now unwanted responsibilities, without, of course, any thought of the damaging effect on the environment. If these wild caught birds have not been held in captivity for more than several months and they are still reasonably healthy, a large percentage will establish back into the wild. This situation is exactly what happened with the Eastern Long-billed Corella in the early 1980's, when many thousands were trapped legally in Victoria and sold legally into the pet trade of New South Wales and to date the environmental consequences have been totally ignored by all wildlife authorities.

My first sighting of this species many years ago was of six birds feeding on the ground in western Victoria. In those days I considered myself lucky to see these few birds in the wild at a time they were protected and considered rare fauna in New South Wales. More recently I have seen flocks of hundreds feeding on both green and dry pastures throughout western Victoria. This reappearance of the Long-billed Corella in localised areas must not lull us into a false sense of security. If not closely monitored, the current control measures including legal trapping and illegal poisoning could see the demise of this species this century.

This species inhabits open grasslands and farmlands usually close to treelined watercourses and is seldom found far from permanent water. They are often seen feeding in the company of Sulphur-crested Cockatoos and Galahs. After feeding on the open grasslands or farm land this cockatoo returns to tall stands of Eucalypt to rest during the heat of the day and to roost at night.

Localised movements and population fluctuations that have been observed in some areas, particularly after breeding, are probably only related to food supply and weather conditions. Congregation into flocks in excess of a thousand birds is often seen during the summer months in areas where food is plentiful, following a good breeding season.

The diet consists of the seeds, roots and bulbs of various native and introduced grasses and ground plants of which the introduced onion grass *Ramulea longifolia* is a favourite as well as various nuts, fruits and berries. Crop damage, particularly in grain growing areas has led to extensive persecution of this species in Victoria.

Although sympathy must be felt for the farmers whose livelihood is jeopardised through the crop damage caused by this species, surely the survival of the species must be of utmost importance.

Surely the survival of ANY species must be of the utmost importance to the Australian nation or for that matter the world. The relocation of the affected farmers or subsidies for loss of income caused by the birds, must be preferable to the destruction of a species.

The flight of this cockatoo consists of a number of medium-length wing beats followed by varying short periods of gliding flight. It is similar to the flight of the Sulphur-crest but with a more pronounced wing beat.

This species has often been confused with other Corella species in the wild. The long narrow bill is not always identifiable when viewed from a distance. The most obvious identifying feature is the orange red band across the throat that occurs only in the Eastern Long-billed Corella.

Breeding in the Wild

The preferred nesting site is a large hollow limb or hole in the trunk of a tall living Eucalypt, usually in the vicinity of water and often in a high inaccessible position more than 15 m (50 ft) above the ground. Breeding activities usually commence in July and may extend to November. It involves the male and female preparing the hollow with minor chewing and excavation within the log, amid much noise and courtship display.

This cockatoo has been known to burrow into cliffs and high banks to nest, when suitable hollows were not available.

Two or three oval white eggs are laid on a base of chewed rotten wood dust. Incubation is shared by the male and female, with most of the daily duties being carried out by the male. The female does the late afternoon, nightly and early morning incubation.

The chicks hatch with sparse, whispy off-white down and are fed by both parents. Fledging occurs approximately eight weeks after hatching and the parents continue to feed their constantly begging young for many weeks.

The formation of family parties and flocks occur as the young near independence and move further afield in search of food.

Aviculture

Until comparatively recent times this delightful cockatoo whose extreme ugliness, I have always felt is its beauty, was kept almost totally as a caged pet in Australia

Then almost 20 years ago its rarity in aviaries was recognised and many aviculturists made serious attempts to breed them and many successful breedings occurred. Then the avicultural world was flooded with 10,000 wild caught Long-bills. Aviary bred specimens became difficult to dispose of thus making this cockatoo less desirable and the number of pairs held in aviaries for breeding purposes was drastically reduced.

Prestwich, in his "Records of Parrots Bred in Captivity" Part 2, records with some reservation the supposed breeding of this species by Prince Demidoff of San Donato, Italy, prior to 1854. The record is without data and must be considered doubtful. Over 100 years pass before another reference is made to the breeding of the Eastern Long-billed Corella in avicultural literature. The Curator of Birds in San Diego Zoo, USA, K. C. Lint, records the rearing of one young Eastern Long-billed Corella in 1959 and again in 1960. In my opinion this must be considered the world's first breeding for this species.

The next recorded success was by Mrs C. Pullen of Wellington, New Zealand, whose breeding pair reared eight young in four seasons from 1965 to 1968.

In Australia the first recorded breeding was by Alan Lendon in 1968. The following is an abbreviated account of this breeding taken from his revised edition of Cayley's "Australian Parrots in Field and Aviary".

The pair, who were reputed to have laid eggs the previous season, were assessed by Alan Lendon as an unlikely looking breeding pair, the male being timid and unable to fly because of a damaged wing and the female a hand reared bird. They were housed in an aviary approximately 4 m (13 ft) long, 1 m (3 ft 3 in.) wide and 2.4 m (8 ft) high and supplied with a log about 1.5 m (5 ft) long and 15 cm (6 in.) internal diameter and hung obliquely in the shelter of the aviary, with a perch affording easy access for the male.

Immediate acceptance of the log was obvious and mating was soon observed. The first egg was noticed in the log on 22 September, 1968, and a second egg the next day followed by a third egg four days later. Incubation commenced with the laying of the second egg and was carried out by the male and female, with the male sharing the duty during the day. Two of the eggs were fertile with the first hatching on 16 November, 1968, and the second egg hatching the following day with the probable incubation period of 24 days. One of the clutch, probably the youngest, died about one week after hatching, but the other flourished and left the nest on 10 December, 1968, 55 days after hatching.

The diet consisted of sunflower seed, hulled oats and canary seed together with nut grass, milk thistle and seeding grasses.

My own first breeding success with this species occurred in 1984 after attempting with the pair for 13 years. I have always found patience to be an asset when breeding cockatoos. The pair consists of an old hand reared cock, who says a few words and who is in excess of 35 years old, with a younger probably wild caught hen. They appear to be totally incompatible.

Each year this pair would excavate a huge hole under the concrete floor of the house section of their aviary and totally ignored every type of hollow log I provided over the years. On a number of occasions I was reasonably sure eggs had been laid, although I could never verify this due to the length of the excavated tunnel. Every year the breeding attempt would be aborted because of cave-ins, flooding rains or excessive heat under the exposed concrete floor.

After yet another failure in 1983 I decided to have both birds surgically sexed, just in case they weren't a pair. They proved to be a pair.

During the winter of 1984 I moved the pair to another aviary which is 7.3 m (24 ft) long, 1.2 m (4 ft) wide and 2.15 m (7 ft) high. The aviary has solid walls for its entire length and a 2 m shelter on each end. The top in the centre is left open as well as the front which faces north. There was a derelict log hanging vertically in the shelter, which was 75 cm (2 ft 6 in.) long and 30 cm (12 in.) in diameter, with an opening in the side at the top and a fitted metal lid on the top of the log. My intention was to replace this old log with what I considered a more suitable nest site, but this pair of Long-bills adored this old log from the first day.

A few weeks prior to breeding is the only time any affection is shown by this male towards the female. Even during incubation the male will often attack the female without provacation. This is not normal behaviour, almost all other pairs that I am familiar with are quite devoted.

The first egg was laid on 5 September, 1984, followed by a second egg three days later. Incubation was shared in the normal manner although unfortunately one egg was broken during incubation. The remaining egg hatched 24 days after incubation had commenced. The chick hatched with a sparse silvery off-white down and developed slowly over the first two weeks, but then grew rapidly and fledged 55 days after hatching. This remarkably is the same fledging period recorded by Alan Lendon.

During the rearing period the pair were fed a basic diet of sunflower seed together with sprouted sunflower, sprouted corn (maize), sprouted French white millet, raw peanuts, green peas in the pod, silverbeet, apple and wholemeal bread which they particularly relished. They were fed whatever quantities of each item that they would consume in a day.

Sexing

Visual sexing of this species is difficult to say the least. Some females have a slightly smaller head and mandible and others have slightly less orange red on the face and throat. No doubt a lot of time can be saved by having these cockatoos surgically sexed. Immature birds are also difficult to sex visually.

Display

The male approaches the female in a strutting motion, with the crest erect, while uttering an unusual guttural yodelling sound unlike the call of any other Australian cockatoo.

Nests

I have found Eastern Long-bills to be more difficult to please with nesting sites than any other cockatoo. Rather than suggest a particular type of log or box I would advise a variety of nest sites. Large and small logs, horizontal, inclined and vertical, on the floor of the aviary and hung at various heights. Then hopefully the particular needs of your pair of birds will be achieved. I have used natural wood dirt, leaf mould, decaying lawn cuttings, dirt, peat moss and various mixtures of these ingredients, as filling for the hollows in my efforts to have a nest accepted. When my success was finally achieved, fine wood shavings were used in the log.

Nesting and Hatching

Breeding activities usually commence with noisy courtship displays and increasing interest in the nest site which is frequented by both birds who usually excavate some of the filling. Should too much be removed exposing the bare base of the log, more must be added as eggs laid on the

A Nestling one day old.

B Nestling two weeks old.

C Nestling three weeks old.

D Nestling 4-and-a-half weeks old.

E Nestling six weeks old, just prior to fledging.

Plate 33. Eastern Long-billed Corella — Nestling.

bare base of the nest site are often cracked or broken. Long-bills do little chewing around their nest, or to their perches for that matter, in comparison to other cockatoo species.

Long-billed Corellas in the Sydney region usually breed in the spring. My records show that eggs have been laid from late August to early October. A clutch of two or three and very occasionally four oval white eggs laid at two or usually three-day intervals. Occasionally longer periods between eggs have been recorded. We have recorded fresh egg weights of 18.28 g and 20.68 g.

Incubation may commence with the laying of the first egg but usually starts when the second egg is laid. Incubation periods of 24 and 25 days have been recorded. The chicks hatch with a wispy off-white down and horn coloured beaks. They develop quite slowly at first. Eyes start to open at two weeks of age and pin feather development is visible soon after; then at three weeks crest feathers can be detected. At four weeks chicks are half feathered and fully feathered at just over six weeks. Fledging usually occurs between seven and eight weeks of age. Immatures are similar to adults except for shorter upper mandible as well as less and duller orange red colouring on the face and throat.

Both parents feed the young in the nest and continue to do so for some weeks after fledging. Often the male is more attentive to the young after they leave the nest than the female.

Young birds are usually independent four weeks after fledging and can be safely removed from their parents about 10 weeks after leaving the nest.

Mutations and Hybrids

I can find no records of a mutation occurring in this species but hybrids have been recorded with the Greater Sulphur-crested Cockatoo and the Galah.

GENUS NYMPHICUS

Plate 34: Cockatiel *Nymphicus hollandicus* — Adult pair.

COCKATIEL

(Nymphicus hollandicus)

Derivation:

Nymphicus — *Nympha*, Latin for nymph
icus, Latin for having the nature of

hollandicus — Latin, taken from New Holland

Description:

Length 31 cm average
Weight 91 g average
Appearance as in Plate No. 34

Classification

The Cockatiel, *Nymphicus hollandicus*, is the sole member of the genus Nymphicus which in turn is the only genus of the second Cacatuinae group, Calopsittacini. Although some slight colour variations occur throughout the extensive range, no subspecies have been recognised.

After many decades of indecision by taxonomists the Cockatiel eventually, and rightly so I believe, came to rest amongst the Cockatoos. The general body shape is more parakeet-like than that of a cockatoo, but the following features and habits confirm the close affiliation with the cockatoos.

The relationship between Cockatiels and cockatoos is probably more noticeable during the first few weeks of their life than at any other time. The similarity of mandible formation, feeding attitudes and general body shape is most prominent at this stage.

A movable crest is typical of all cockatoos including the Cockatiel.

Male and female share incubation duties similar to most cockatoos and unlike all parakeets.

When nesting, Cockatiels usually back down the nesting hollow similar to many cockatoo species.

Both Cockatiels and all other species of cockatoos have powder down patches in their plummage.

Earliest Report

The Cockatiel was recorded during Cook's voyage of discovery to Australia and was described by Latham in 1781 with the name of Crested Parakeet, although without any locality information. Later Gmelin gave it the Latin name of *Leptolophus novaehollandiea*, which was accepted and used by Latham in his "Index of Birds". However, Kerr discovered Gmelin had previously used the name *novaehollandiea*, so he altered the name to *Psittacus hollandicus* in his "Animal Kingdom" Volume 1, part 2, page 590, 1790.

The genus name of Nymphicus was adopted from Wagler, Abhandl. Ak. Wissen. Munich, Volume 1 page 490, 1832.

Apparently the first illustration was in Swainson's "Zoology Illustrated" Series 2, Plate 112, 1832, under the name of *Leptolophus auricomis*. Lear also illustrated this species in 1832 in his "Illustrations of the Family Psittacidae" Plate 2 as *Palacornis Novae Hollandiea.*

Range, Habitat and Field Notes

The Cockatiel ranges over most of Australia, being absent only from Cape York Peninsula, the south-western coastal area, Tasmania and the arid deserts of Western Australia. However, I would suggest the occurrence of this species in these desert regions after rain would be a strong possibility.

Data collected by the RAOU for their "Atlas of Australian Birds" suggests generalised population movements southward in the summer and northward in the winter. Movements and population densities are also influenced by good seasons and the occurrence of heavy falls of rain.

Although generally recognised as a nomadic species, resident populations, whose numbers fluctuate according to the season and hence the availability of food do occur in the southern section of the range. This is a particularly successful species whose numbers appear to have increased since early white settlement, due, no doubt, to the establishment of permanent drinking water for stock and the creation of grasslands through the clearing of trees and scrub. These grasslands suit their ground feeding habits.

I have been unable personally to substantiate the persistent rumours of an established breeding colony of Cockatiels in the rural suburbs south-west of Sydney, although many eyewitness reports appear to confirm their existence.

My first field sighting occurred near Brewarrina, New South Wales, in 1961, when a flock of 100 or so birds were sighted, feeding on dry open grassland. My notes taken at the time suggest I was impressed by the rapid direct flight when flushed as well as the turning and wheeling of the flock in unison, similar to a flock of pigeons.

The most memorable of my numerous sightings of this species in the wild was during a trip across the plains of central Queensland in May 1971. Flocks of about 100 Cockatiels started to appear, feeding along the sides of the roadway in the vicinity of Blackhall. For the next 650 kilometres of the journey to Cloncurry there was barely a kilometre travelled that a flock of similar proportions was not encountered. The flocks would rise as our vehicle sped by, uttering their distinctive two-syllable whistling call whilst wheeling around before alighting on the ground to feed again or occasionally covering a small shrub or dead tree on these almost treeless plains.

This concentration of Cockatiels across the vast grassland plains was due to falls of rain throughout the region some weeks earlier, which provided the growth that these birds were now harvesting the half ripe seed from. The prevalence of the flocks close to the roadway was due to the lush growth created from the additional water shed off the road surface to the adjacent areas.

When passing through the same area the following August all but a few Cockatiels had gone from these plains but upon arrival in the hilly country of the Cloncurry-Mt Isa region which is criss-crossed with tree-lined dry water courses, I found huge numbers of Cockatiels had moved from the plains into this area to breed. Every conceivable hollow had Cockatiels attempting to nest in them. I am sure the breeding success rate in that area that year was quite low due to the high concentration of birds all competing for food in an area that, even at the best of times, is quite arid.

Obviously, as reflected by their huge range, Cockatiels inhabit a large variety of habitat, although a preference for the inland open forest, grasslands and treelined watercourses in open country is apparent.

Their diet consists of seeds, fruits and berries of grasses, trees and shrubs, with acacia seeds being particularly relished. Crop damage occurs in grain growing areas and occassionally in orchards. This never reaches the extent of that inflicted by some of the larger species of cockatoos.

Flight is fast and direct and typified by the regular beats of the long, pointed, swept back wings. The white wing patches are visible on the top side of the wing as the flocks wheel in flight. A resemblance to the Polytelis genus is said to exist when the slender form of the Cockatiel is viewed in flight.

Breeding in the Wild

Cockatiels usually select a hollow limb or hole in a tree, which is large and roomy, unlike other Australian psittacines who normally prefer a hollow which appears a little too small for their needs. The hollow may be in a living or dead tree, often with a dead spout entering a hollow in a living Eucaypt being preferred. Nest site heights of under 1 m (3 ft) to in excess of 8 m (26 ft) have been recorded. The depths of the hollows vary considerably but average about 50 cm (20 in.).

Breeding usually occurs from August to December in the southern parts of the range, but in the arid interior and in the north it is erratic, being influenced by rainfall hence breeding takes place when conditions are suitable. Records of autumn, winter and spring breedings are common throughout the central and northern sector of its range.

Three to eight white oval eggs are laid on a base of wood dirt or rotten wood and are incubated by both male and female for a period of 19 to 21 days. Chicks hatch with yellow down and are fed in the nest by both parents. They develop quickly and usually fledge about five weeks after hatching. After fledging family parties, small flocks and occasionally large congregations are formed while nomadically foraging for food.

Aviculture

The Cockatiel appears to have been first bred in captivity in France prior to 1850. Dr Carl Russ in his "Parrot" volume, page 688, refers to a breeding in the "Jardin des Plantes" during 1846 and later in his 1901 edition, page 303, states this species has been bred since 1846. Russ also refers to Leuchfeld of Nordhausen's success in 1858 as the first German breeding and this provides the first detailed breeding account to be published. This species bred prolifically throughout Europe and England during the latter half of the nineteenth century as confirmed in Prestwitch's "Records of Parrots Bred in Captivity".

No doubt the cockatiel now ranks second only to the budgerigar as the most prolific and widely kept Australian Psittacine as a pet or aviary bird in the world today and rivals the Peachfaced Lovebird as the second most popular member of the Parrot family. Large numbers of this species are bred annually on a commercial basis for the pet trade throughout the world.

Early Australian breeding records are sketchy, but G. A. Keartland's success prior to 1914 is quoted in North's "Nests and Eggs of Australian Birds" Volume 3, page 98; "They breed readily in captivity. I placed a pair in an ordinary packing case with a wire netting front and with a hollow log containing a handful of sawdust as a nesting place. They reared three broods in the season, the hen laying soon after the young ones left the nest. They had two broods of five and one of four."

In Campbells "Nests and Eggs of Australian Birds" Volume 2, page 622, 1901, he states: "The late Mr T. Augustus Forbes-Leith in his pleasant little treatise on "Parrots of Victoria" mentions that he knows of several instances of this parrot breeding in aviaries and rearing young."

As is often the case with prolific species, early breeding successes pass unrecorded because we as aviculturists place little importance on breeding common species.

I first bred this species during the late 1950's but kept no records of the event. The importance of keeping detailed records of all breedings, whether common or rare, must be stressed for no matter how insignificant a minor breeding detail may appear to us now, perhaps this minor detail may be a key to some significant avicultural problem 100 years from now.

My own interest in the specialised breeding of Cockatiels occurred during the late 1960's and early 1970's when I became involved in the establishment of the Lutino Cockatiel in Australia. The aviaries I found to be most suitable for this were small fully roofed and solid partitioned structures. They were open only at the front and measured 2.15 m (7 ft) long, 90 cm (3 ft) wide and 2.15 m (7 ft) high and they faced north. A perch was fitted close to each end to encourage maximum flying. In keeping with the cockatiel's preference for larger nest sites, nest boxes measuring 30 cm (12 in.) cubed with a removeable lid to facilitate cleaning and a 7.5 cm (3 in.) entrance hole and perch near the top of the box were provided. The box was filled to a depth of 2.5 cm (1 in.) with fine wood shavings.

A basic diet of French white millet was fed, together with a daily ration of sprouted sunflower seed, green peas, apple and silverbeet. The supplements to the millet were limited when the birds were not breeding so as to avoid obesity; they were increased gradually as the birds came into breeding condition and fed on an unlimited basis when young were being reared.

Cockatiels housed a pair to an aviary under these conditions proved to be prolific breeders.

This cockatoo has been bred under every conceivable set of conditions, from breeding cabinets and suspended wire aviaries to various sized conventional aviaries and large planted aviaries; housed in pairs, large colonies and with mixed collections of parrots and other birds. When I first met Bob Lynn many years ago, he was breeding from a pair of Cockatiels housed in a large aviary with a pair of Yellow-tailed Black Cockatoos.

Sexing

The adult male is uniform grey in colour with a white wing bar, bright yellow face and orange-red ear coverts. The female differs from the male in having the grey lower chest and underparts barred with the palest yellow. It only has a yellow suffusion in the facial area and dull orange-red ear coverts. The female also has yellowish white barring on the tail feathers, with the exception of the outer feathers which are light yellow barred with grey. The female's flight feathers carry yellowish white spots on the inner web which are visible from the underside when the wing is fanned.

Immatures are similar to the adult female, with some but not all young males showing brighter yellow on the face as well as brighter orange red ear coverts. Young males have usually developed the brighter yellow face by five or six months of age but retain the barred tail feathers and underwing spots until their first moult.

Display

Male Cockatiels seldom display but on rare occasions approach the female with a hopping and bobbing motion, while uttering a low whistling call. I have never witnessed courtship feeding although mutual preening is often seen.

Nests

Hollow logs with an internal diameter of approximately 30 cm (1 ft) and a depth of 45 cm (18 in.), hung in a vertical or partially inclined position, are usually accepted. The larger type vertical or horizontal nest boxes, filled to a depth of about 7 cm (3 in.) with fine wood shavings provide easy-to-service nesting facilities. If a vertical nest box is in excess of 30 cm (1 ft) deep an internal wire ladder should be provided. Any nest site provided should have easy access for cleaning as Cockatiels are particularly dirty nesters and cleaning is often necessary during the rearing period or at least after each brood has fledged.

Nesting and Hatching

The male and female frequent the nest site prior to breeding. Three to eight eggs, and sometimes even more, are laid usually at two-day intervals, although four to six eggs can be considered a normal clutch. Incubation usually commences with the laying of the second or third egg, although this is variable. The incubation is carried out by both the male and the female, with the hen incubating at night as well as sharing the duties with the male during the day. Incubation periods of 19 to 21 days have been recorded in my aviaries. The chicks hatch with a yellow down, pale horn coloured beak and black eye spots. At 10 days of age the eyes have commenced to open and early pin feather development can be seen. A few days later pronounced crest feathers are observed. When three weeks old, pin feathers are well advanced and by four weeks the birds are fully feathered. Fledging occurs about five weeks of age. The young are fed by both parents for two or three weeks after fledging and can be safely removed from the parents three weeks after leaving the nest.

Mutations

I do not intend to delve too deeply into Cockatiel mutations, but shall list the known basic mutations, their history where possible, and their genetic inheritance.

Until 1950, mutations in this species were virtually unknown, except for the occasional record of a mutant being sighted in the field. Since then new mutations have been occurring with such regularity that the Cockatiel must now be considered the third most prolific mutation producing Australian psittacine. The Budgerigar is obviously first amongst all psittacines and the Red-rumped Parrakeet is second since its current explosion of mutations in Australia.

Basic Mutations

PIED

The establishment of a pied mutation by Mr D. Putman of San Diego, California, USA, was reported in 1951. This pied mutation is a dominant inheritance with a visual single and double factor. (See Plate No. 36C.)

LUTINO

In 1958 Mr Cliff Barringer of Florida, USA, bred the first Lutino Cockatiel from a pair of normal appearance. This stock was eventually obtained by Mrs E. L. Moon who was instrumental in fully establishing

A Newly hatched chicks.

B Nestlings two weeks old.

C Nestlings three weeks old.

D Nestlings four weeks old.

Plate 35: Cockatiel — Nestlings.

this glorious sex linked mutation. She marketed them as Moonbeams. (See Plate No. 36A.)

PEARL

The Pearl Cockatiel, which was first bred in West Germany in 1967, is of a typical sex linked inheritance. George Smith suggests that this colour form is in fact an opaline mutation. I cannot accept this theory for in all other known opaline mutations that have occurred in various species of parrots both male and female are affected. In the Cockatiel this mutation occurs in females and immature males but moults out in adult males.

My interpretation of the Pearl mutation is that it is an extension of the pale yellow barring and spotting present in the adult female and immatures which accounts for the young males moulting out the Pearl plummage on the adult moult. (See Plate No. 36B.)

CINNAMON (sex linked)

This cinnamon mutation appears to have been first bred in a Belgium aviary in 1968. It is a typical sex linked cinnamon mutation where the chicks hatch with red eyes but the eyes appear darker by the time they fledge. (See Plate No. 36D.)

AUSTRALIAN RED-EYED CINNAMON (Fallow)

This mutation first appeared in South Australia during the sixties and was partially developed by Fred Lewitska, then Joe Mattinson of New South Wales and finally by Ron Rodda of South Australia who is still breeding this recessive mutation and supplied me with the following information. Ron believes there is a lethal gene involved which inflicts extensive losses on chicks while hatching or soon after. There are two colour phases, a light cinnamon or fawn colour and a silver colour which are usually males.

RED-EYED SILVER

This is a recent mutation to occur in England and Europe which is apparently a very advanced cinnamon (fallow) mutation and is of recessive inheritance.

BLACK-EYED WHITE

Somewhere along the way there appeared a black-eyed white mutation whose origin is obscure and which was often confused with and interbred with the Lutino. This recessive mutation occurred overseas as well as in Australia. I first saw the Australian version of this mutation many years ago in Alan Lynch's aviaries in Queensland and he supplied the following information. The mutation was developed from a wild caught bird and had apparently been bred with a Lutino in the early stages of establishment for occasionally a Lutino would occur amongst the young. Males are an off-white colour with the tail feathers light brown on top and brown underneath. The females were a dirty white colour with variable amounts of yellow flecking in the chest and belly. Alan hasn't any specimens of this mutation left in his collection but believes some still exist in New South Wales.

WHITE-FACED (Charcoal)

This mutation, which is devoid of all yellow colour pigment, apparently first appeared in Holland about 1976. The occurrence of this recessive mutation obviously heralded the eventual production of the true albino, when combined with the Lutino. (See Plate No. 36E.)

AUSTRALIAN CINNAMONS

The first typical cinnamon Cockatiel in Australia which I am familar with was a young wild caught bird which was offered for sale in Sydney in the mid 1970's. The bird was purchased by Frank Parmenter of Sydney who eventually reproduced this mutation. Unfortunately this sex linked cinnamon variety has not been a prolific strain and has not really established itself in eastern Australia.

More recently another cinnamon mutation has been bred in Western Australia from a Lutino hen and a normal coloured cock. In 1985 Jeanette Hickford of Kalgoorlie, Western Australia, obtained a young hen from the original source and has had remarkable success in developing this sex linked mutation.

Both of these Australian cinnamon mutations appear to be distinct from each other as well as being distinct from the cinnamon mutation developed in Europe.

AUSTRALIAN FALLOW

This is another distinct advanced cinnamon mutation which retains its red eyes after maturity. It appeared in a Sydney aviary in 1987, the product of a Lutino hen by a split Lutino male mating. As yet the genetic inheritance has not been determined.

DILUTE OR BLACK-EYED SILVER

Yet another new mutation has appeared in England. It is a black-eyed diluted cockatiel which is of dominant inheritance. It has visual single and double factor forms. (See Plate No. 36F.)

The foregoing list is of known basic cockatiel mutations. When they are mated together they are capable in the subsequent generations of producing combination mutations of any two of these basic colours such as Pied-Pearls, Cinnamon Pied, Pearl-Lutinos, Albinos, etc. (See Plate Nos. 37A, B and C.)

Conglomerate mutation is a term I use to describe a colour form in which three or more distinct mutations have been used in its development, such as Cinnamon-Pied-Pearl. (See Plate No. 37D, E and F.)

Hybrids

I have difficulty in accepting the possibility of the Cockatiel, who is the sole member of such a distinct and isolated genus as Nymphicus, producing hybrids with any parakeet. Nevertheless, the following hybrids have been recorded, some with documentation.

COCKATIEL×RED-RUMPED PARAKEET

Recorded by Dr Green in his "Parrots in Captivity" Volume 1, page 38, 1884, where he states that he was shown "A curious creature said to be a hybrid between a female Red-rump Parakeet and a male Cockatiel which more closely resembled a Cockatiel than a Red-rump."

COCKATIEL×BLUE-WINGED PARAKEET

Don Martin of Victoria, Australia, reported in litt. 5 December, 1948, that a male Cockatiel mated to a Blue-winged Parakeet who laid five eggs, hatched four chicks and reared three — one male and two females. Recorded in the Avicultural Magazine 1949, page 32. (See the Duke of Bedford's comments in the Avicultural Magazine 1950, page 190.)

COCKATIEL×EASTERN ROSELLA

Recorded by Prestwitch in his "Records of Parrots Bred in Captivity" without documentation.

A Lutino Cockatiel — Male.

B Pearl Cockatiel — Female.

C Pied Cockatiel — Male, together with a Cinnamon Pearl.

D Cinnamon Cockatiel — Male.

E White-faced Cockatiel — Male.

F Silver Cockatiel — Male.

Plate 36: Basic Cockatiel Mutations.

no Cockatiel — Male. Combination of
no and White-faced.

B Lutino Pearl Cockatiels.

amon Pied Cockatiel — Immature.

D Primrose Cinnamon Pearl Cockatiel — Immature.

Pearl and Primrose Pied Pearl Cockatiels.

F Cinnamon Lacewing White-face Cockatiel — Hen.

Plate 37: Combination and Conglomerate Cockatiel Mutations.

REFERENCES

J. M.Forshaw — Australian Parrots. 2nd ed. Landsdowne Press.

J. M. Forshaw — Parrots of the World. 2nd ed. Landsdowne Press.

RAOU — The Atlas of Australian Birds. Melbourne University Press.

A. Lendon — Neville W. Cayley's Australian Parrots in Field and Aviary. Angus & Robertson.

Australian Aviculture — Avicultural Society of Australia, c/- 52 Harris Road, Ellminyt, Victoria 3249.

Bird Keeping in Australia — Avicultural Society of South Australia, P.O. Box 4, Blackwood, South Australia 5051.

The Emu and Newsletter of the Royal Australian Ornithological Union, 21 Gladstone Street, Moonee Ponds, Victoria 3039.

The Australian Birdwatcher — Journal of the Bird Observers Club of Australia, P.O. Box 185, Nunawading, Victoria 3131.